WHAT **BUGGED** THE
# DINOSAURS?

# WHAT **BUGGED** THE
# DINOSAURS?

*Insects, Disease, and Death in the Cretaceous*

GEORGE POINAR, JR.

AND ROBERTA POINAR

*WITH PHOTOGRAPHS AND DRAWINGS*

*BY THE AUTHORS*

PRINCETON UNIVERSITY PRESS

PRINCETON AND OXFORD

Copyright © 2008 by Princeton University Press
Published by Princeton University Press, 41 William Street, Princeton,
New Jersey 08540
In the United Kingdom: Princeton University Press, 3 Market Place, Woodstock,
Oxfordshire OX20 1SY

*Library of Congress Cataloging-in-Publication Data*

Poinar, George O.
    What bugged the dinosaurs? : insects, disease, and death in the Cretaceous /
George Poinar, Jr. and Roberta Poinar ; with photographs and drawings by
the authors.
        p.   cm.
    Includes bibliographical references and index.
    ISBN-13:  978-0-691-12431-5 (alk. paper)
    1. Paleoecology—Cenozoic   2. Dinosaurs—Diseases.   3. Paleoecology.
4.   Insects, Fossil.   5. Insects—Ecology   I. Poinar, Roberta.   II. Title.

QE720.P65 2007
560′.45–dc22        2007061024

British Library Cataloging-in-Publication Data is available

This book has been composed in Palatino
Printed on acid-free paper. ∞
press.princeton.edu

Printed in the United States of America

10   9   8   7   6   5   4   3   2   1

*This book is dedicated
to the inquiring minds of
future generations.*

*So, naturalists observe, a flea*
*Hath smaller that on him prey;*
*And these have smaller still to bite 'em;*
*And so proceed* ad infinitum.
*Thus every poet, in his kind,*
*Is bit by him that comes behind.*
—*Jonathan Swift (1667–1745)*

# Contents

# Preface

ONE HUNDRED MILLION years ago, dinosaurs ruled the earth . . . or did they? In reality, there were millions of tiny animals undaunted by those powerful reptilian behemoths and unfazed by their reign of terror, that actually sought *them* as prey. Hordes of belligerent biting insects assaulted the majestic *Tyrannosaurus rex* much the same way that they pester humans now. During the Cretaceous period, insect populations, unchecked by insecticides as they are today, thrived and undoubtedly accounted for the majority of animal diversity and biomass on the earth. With nearly a million species of insects described and possibly three times as many still unidentified,[1] ours is clearly an insect world. Imagine what it was like 100 million years ago when insect diversity was even greater, and consider that maybe, just maybe, it was the insects that ruled the world. And if you are not convinced of the ultimate superiority of the insects over dinosaurs, just consider this: insects were around before, during, and after the reign of the non-avian dinosaurs.

We would like to take you on a journey through time to examine the world of the dinosaurs and discover what bugged them. By using insect fossils from the Cretaceous period, we'll visualize the likely relationships that occurred between insects and dinosaurs, and try to predict how they could have impacted dinosaur populations. Our interpretations of the habits of the fossil insects will be based on the behavior and ecology of their present-day descendants.[2] Crucial to this endeavor are several important amber deposits that provide glimpses of insects that shaped the environment at three important periods: Early Cretaceous Lebanese deposits dating from 130–135 million years ago (mya), mid-Cretaceous Burmese deposits of some 99–105 mya, and Late Cretaceous Canadian deposits of 77–79 mya. Other

Mesozoic fossils will also contribute to our view of that vanished world. Finally, we will examine the hypothesis that insects vectoring disease-causing agents could have contributed to the decimation of already threatened dinosaur populations and led to at least local and even global extinction.

Studies of the past are frequently infused with controversy. This book is not about debating unresolved issues, taking sides, or creating new ones. For our study of the role of insects in the Cretaceous, it doesn't matter whether dinosaurs were cold- or warm-blooded or in between or all of the above. It's not significant if some dinosaurs were the ancestors of birds. It wouldn't really matter if some of the fossils were incorrectly identified or if the present theories on extinctions at the end of the Cretaceous are being challenged. This is a story about visualizing an ancient web of life. The *T. rex* and sauropod plucked from the past and plastered across the big screen were not isolated entities. They were part of a complex that was inextricably interwoven with all the other faunas and floras in their habitat, and in a very real sense with the entire planet. To illustrate this, some chapters begin with a scene depicting how we envision life in the Cretaceous. These are printed in a different font.

Our idea is to paint in some of the basic structure of that past web. Others will contribute fine interconnecting pieces; some of what we say will be deemed erroneous and discarded, and some will fit and remain. The canvas is essentially just being roughed in for future generations to work on, but it is unlikely that the picture will ever be completed. The world today is like an ever-changing, pulsating mega-organism, the complexity of which is just beginning to be understood even as man destroys it. The ecology of the Cretaceous world would have been certainly even more complex than that of today, with more floral and faunal diversity. Many of those plants and animals didn't survive to the end of the Cretaceous. The basis for their extinction will ultimately be found in a massive disturbance of global ecology—a tear in the web of life.

# Acknowledgments

WE WOULD like to thank the following people for various types of assistance provided during the preparation of this work: John Aitchison, Norm Anderson, Nick Arnold, Dennis Braman, Kenton Chambers, Peter Cranston, Bryan Danforth, Jim Davis, Jane Gray, Penny Gullan, Grand Huang, R. L. Jacobson, Alexander Kirejtshuk, Bob Mason, Raif Milki, Cheryle O'Donnell, Ted Pike, Barry Roth, John Ruben, Ryszard Szadziewski, Sam Telford Jr., Jack Tkach, and Charles Wellman. Special thanks are extended to Alex Brown for assistance in locating amber specimens, Ron Buckley and Scott Anderson for loaning specimens from their Burmese amber collections, and Art Boucot for stimulating discussions. All specimens illustrated here are deposited in the Poinar amber collection maintained at Oregon State University except for the following: specimen in color plate 6A is from the Scott Anderson collection; specimens in color plates 5E, 8A, 12D, 13C, and 15A are from the Ron Buckley collection; specimens in color plates 2C, 5A, 5B, 6B, 6D, 7C, 9B, 11D, 12B, 13A, 13B, and 15B are from the Royal Tyrrell Museum, Alberta, Canada; and the flower in color plate 14A is from the Grand Huang collection. Design templates are by Greg Poinar.

WHAT **BUGGED** THE
# DINOSAURS?

# Introduction

The surviving members of a herd of ornithopod dinosaurs grazed along the edge of a Cretaceous conifer forest. It had been a particularly hard and long dry season. A few individuals fed sporadically on the parched sedges and horsetails growing near the banks of a meandering river. The river had dwindled down to a trickling stream flowing between high-cut banks. Once numbering in the hundreds, disease had now reduced the herd to less than fifty—the very young and the aged were conspicuously absent. Many of the majestic animals appeared lethargic and even were oblivious to the pack of predatory theropod dinosaurs that followed. Those fierce carnivores, armed with sharp teeth and sickle-like claws, were fat and satiated because sick herbivores made easy prey.

Ordinarily, these plant-feeding dinosaurs spent at least half of their waking hours chomping on the tender vegetation sprouting around streams or in open meadows. However, with waning appetites, the infected animals had not fed for days and stumbled along in a debilitated state. They normally avoided the direct rays of the afternoon sun, but now many stood motionless in its intense heat. Frequently they would shuffle down to the waters edge, laboriously bend down, and drink for long periods, apparently forgetful of the dangers posed by lurking crocodiles. Persistent diarrhea had dehydrated them and their thirst was almost insatiable. The surrounding terrain was discolored by bloody stools that attracted hordes of flies and beetles.

One trembling ornithopod, with dry skin clinging to prominent ribs and vertebrae, staggered off to one side and began to vomit strands of bloodstained mucus filled with glistening, writhing roundworms. With eyes now reduced to narrow slits, the sick individual was too exhausted to dislodge the ravenous masses of annoying *insects crawling over his thin scaly skin while seeking sites to engorge themselves. When the dying animal finally collapsed, a few members of the herd came over and nudged him, but there was no response and they moved off, giving way to the advancing theropods. The carnivores started tearing away at the carcass, not realizing that they were eating infected meat and being attacked by the same insects that had previously fed on the diseased dinosaurs. Several of them, however, were beginning to show the first signs of infection and withdrew from the feeding frenzy to lie down and rest after only a few mouthfuls. As the others were devouring the remains, a contingent of mites and ticks seized the opportunity to move their residence from the corpse onto the skin of the theropods.*

If an autopsy had been made on this ornithopod, it would have revealed many parasites and pathogens inhabiting the tissues. Some, like amoebic dysentery, malaria, and ascarid roundworms, would have caused lesions in the gut, liver abscesses, and distorted blood cells. But the actual cause of our dinosaur's death would have been listed as leishmaniasis, a protozoan disease. Just like the other members of the herd, he was the victim of an emerging pathogen that was decimating the Cretaceous world. Some 100 million years ago, some of these microorganisms developed novel relationships with biting flies, when the flies' previously harmless symbionts turned into deadly pathogens. In an unprecedented alliance, these insect-borne infections together with already long-established parasites became more than the dinosaurs' immune systems could handle. Sweeping epidemics began changing the herbivore-carnivore dinosaur balance that had existed for millennia. Armed with their deadly weapons, biting insects were the top predators in the food chain and could now shape the destiny of the dinosaurs just as they shape our world today.

Even as the remaining members of this herd succumbed to disease, insects were busy ensuring that the epidemic would spread. Biting flies, sucking the blood of the infirm, were collecting pathogens to inject into other victims. Because of their ability to fly, they could disperse and infect other susceptible dinosaurs within their range. Flies, beetles, and cockroaches visiting the infested feces and cadavers picked up bacteria, protozoans, and nematodes that were then carried to contaminate other vertebrates. Dinosaurs that dined on cockroaches now carrying eggs of ascarids would end up with stomach lesions.

On a larger scale, as the outbreak killed off the herbivorous dinosaurs, the balance of their ecosystem was destroyed. Carnivores may have initially benefited because the dead and dying were plentiful. The downside of this apparent bounty was that they too were becoming ill as their food supply was dwindling. In the following months as the entire ornithopod community crashed, they would face starvation. The combination of hunger and multiple infections would hasten their demise. Vegetation the ornithopods normally fed on would flourish, along with any herbivores that also utilized these plants for food. Those specialists dependent on ornithopods for survival would decline. Others would ultimately move into the niches they left vacant and life would go on. Whether populations would eventually recover depended on many factors, but insect-borne diseases were then, and still are, capable of bringing any animal to the brink of extinction.

Insects not only impact the world because of the diseases they transmit, but in innumerable other ways. They may be small but they are the most diverse group of living organisms and probably have been the most significant ecological force on land since they first arose some 350 million years ago. Insects account for over 57% of the diversity of life on earth and 76% of all animal life. Currently there are over 990,000 species known, while many more have yet to be discovered. Comparing their numbers to mammals, which comprise 0.35% of the species, only serves to accentuate their overall importance.[297]

Herbivorous insects, which make up 45% of their total species, represent about one quarter of all living species.[4] Phytophagous forms consume significantly more plant tissue than vertebrates in every biome studied except grasslands. Indications from the fossil record confirm that this has been the case since the first terrestrial ecosystems became established.⁵ It can therefore be assumed that they were serious competitors with herbivorous dinosaurs during the Cretaceous.

About a third of all organisms on the earth are insects with carnivorous and saprophagous food preferences. Just the fact that carnivorous insects represent about 20% of animal species in the biota tells us how important they are in keeping arthropod populations in check. Predatory insects usually feed on other insects, but some have larger prey. For example, horsefly larvae have been known to kill small frogs, and large praying mantids can take down small lizards, birds, and unwary mice.

The carnivorous insects also include parasites that live on the inside or outside of their victims. A good portion subsist on the blood of animals. Collectively, these are the ones that both fascinate and horrify humans. They also instill us with fear because one bite can potentially lead to death. Biting insects transmit viruses, bacteria, protozoa, and nematodes. One of these, mosquito-borne malaria, kills over a million people each year and is the leading cause of death in children under the age of five worldwide. That means about one in fifty-six people dies every year from just this one insect-borne disease. So we furiously swat, squash, screen, and spray trying to avoid these pests. However, they always manage to find us because they have had eons to adapt to feeding on all terrestrial animals. They certainly unmercifully plagued dinosaurs just as they do us.

Normally humans wouldn't think twice about saprophagous insects—those indeterminate legions that devour dead and dying organic matter. But those that eat excrement, carrion, and detritus are a significant and necessary component of our world, constituting about 11% of the biota. They are the cleaners, charged with the task of disposing of the by-products of life, and they recycle waste with an amazing efficiency. Saprophagous insects have always been an integral part of any ecosystem and undoubtedly were as important in the Cretaceous as they are now.

While insects feed on plants and other animals, they are themselves a significant source of nutrition in the food chain. Small and numerous, they occur in all terrestrial habitats. Convenient packages of protein and other essential nutrients that are readily available and comparatively easy to obtain, insects are consumed by a multitude of creatures. In the Cretaceous, they would have been an important part of the diet for dinosaur young as well as the smaller animals dinosaurs consumed.

The minute but mighty insects have exerted a tremendous impact on the entire ecology of the earth, certainly shaping the evolution and causing the extinction of terrestrial organisms. The largest of the land animals, the dinosaurs, would have been locked in a life-or-death struggle with them for survival. Details of this competition can be garnered from the fossil record. Fossils, interpreted by comparison with their modern counterparts,[2] tell us how insects could have impacted dinosaurs and the entire Cretaceous world. Their preserved remains are the basis for a journey that will take us into the past and reveal new facets of dinosaur ecology and demise. Fossils will be the keys we use to unlock the secrets of the Cretaceous.

# 1.

## Fossils: A Time Capsule

*It was not long after the predators finished feasting on the corpse of the diseased ornithopod that the first raindrops of an approaching storm began to fall. Many of the animals paused in their activities to greet the cooling wind, and as the rain began to pelt down in earnest, some may have scampered for cover. The blood-sucking flies that had been engorging on the disease-ravaged ornithopods sought shelter from the encroaching storm and winged their way among the branches of towering Araucaria trees in the nearby forest. Several individuals brushed against glistening deposits of resin and became stuck. They tried desperately to free themselves, but struggling only pulled them deeper into the golden depths, and finally another flow of resin engulfed and sealed them into their amber tomb.*

*Meanwhile on the nearby floodplain, the small herd of ill dinosaurs continued to graze lethargically on the dry mats of vegetation blanketing the banks of the seemingly placid river, blissfully unaware that upstream, storm runoff was coursing down gullies and ravines and rapidly raising the water level to flood stage. The current began to increase and several sick ornithopods on a small midstream island suddenly became aware that water was lapping around their feet and rising quickly. In the now turbulent flow, heading for shore was not an option, and hampered by their weakened state, they were swept away and drowned. Their bodies were borne downstream to a bend in the river where they floated into an eddy and settled to the bottom. Leaves dislodged by the downburst and lifted by the wind fell into the river and drifted down around their bodies. Just as fast as the cloudburst began, it ended.*

*The muddy waters began to flow slower and a layer of sediment settled over the dinosaur bodies together with the leaves and twigs carried in the debris. The sun came out and dried the raindrops as the river subsided back into its normal channel. Insect sounds once again filled the air and life continued.*

These events would be uncovered millions of years later when the victims would be revealed to us as fossils frozen in rock and entombed in amber from an extinct araucarian forest. It is from such fossils that we will attempt to unravel a story of struggle, terror, and disease in the Cretaceous, one that involves insects, dinosaurs, and their food plants.

Many people have asked us why we have such an inordinate fascination with fossils, particularly those in amber. The answer is complex but several reasons stand out. Fossils are intriguing because they put human time into perspective. Few of us ever contemplate the primeval past because we live our lives bound to the present, measuring time with clocks and calendars. Fossils remind us of the irrelevancy of our own ephemeral time. How significant is *Homo sapiens* when our species has only graced this planet for little more than 200,000 years, an infinitesimal blip in the entire chronology of life? The creatures we are studying and writing about existed in the Cretaceous, a spacious interval that began around 145.5 mya (million years ago) and ended about 65.5 mya[6] (fig. 1). Just considering the magnitude of that period of time frees us from a rather limited viewpoint of the importance of our own.

Fossils are a rarity delivered to us against overwhelming odds, and that alone makes them spellbinding. To be able to touch the bones of a colossal carnivore that flourished 100 million years ago or peer into the gut of an insect that may have fed on that dinosaur is a privilege. Dinosaurs were around for an astonishing 165 million years. Just how many of them lived and died in that time—trillions? Any figure we come up with is only a guesstimate. While tens of thousands of dinosaur bone and teeth fragments have been found, relatively few articulated dinosaur skele-

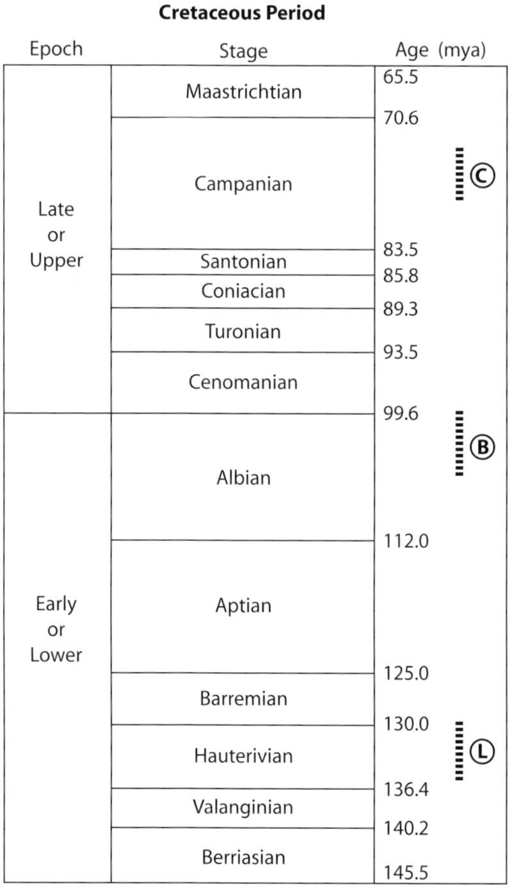

**Cretaceous Period**

| Epoch | Stage | Age (mya) | |
|---|---|---|---|
| Late or Upper | Maastrichtian | 65.5 – 70.6 | |
| | Campanian | | ⓒ |
| | Santonian | 83.5 – 85.8 | |
| | Coniacian | 89.3 | |
| | Turonian | 93.5 | |
| | Cenomanian | 99.6 | |
| Early or Lower | Albian | | ⓑ |
| | | 112.0 | |
| | Aptian | | |
| | | 125.0 | |
| | Barremian | 130.0 | |
| | Hauterivian | | Ⓛ |
| | | 136.4 | |
| | Valanginian | 140.2 | |
| | Berriasian | 145.5 | |

FIGURE 1. A geological time scale showing the Cretaceous Period and its associated epochs and stages along with the approximate ages in millions of years ago (mya).[6] The dotted bars show the estimated times for the Lebanese (L), Burmese (B), and Canadian (C) amber deposits.

tons have been recovered, and just touching one of these is a gift beyond compare. In 2000, the dinosaur expert Dale Russell[7] commented that "although paleontologists have been collecting the remains of Mesozoic animals for more than a century, the total number of known fragments of dinosaur skeletons is only about 5,000". Since then, the number of whole or partial skeletons has certainly increased, but is it any wonder that the auction of "Sue," the skeleton of a *T. rex*, brought in nearly $8.4 million—a veritable bargain considering.[8]

While these reasons make working with rare fossils a truly wonderful experience, a great part of their fascination lies in the stories they reveal about the past. When thinking of insect fossils, most people naturally envision amber, although some wonderful rock impressions also exist. For us, amber represents the crowning jewel of all fossils because of the unique lifelike appearance of the entombed life forms and its ability to capture extremely small and delicate objects like spider webs and microbes (fig. 2). Amber also transforms insects into gems of breathtaking beauty. Magnified by a microscope and illuminated by an amber glow, the backs of metallic beetles ornamented with striations and pitted with craters become sculptures. The multifaceted iridescent eyes of horseflies peer eternally into the surrounding resin, and one can only imagine what scenes they looked upon millions of years ago. Amber often captures the final throes of life, such as female insects desperately trying to ensure future generations by laying eggs in their amber coffins.

But the ultimate thrill of working with fossils in amber lies in the fact that each piece of amber potentially has secrets to reveal. Victor Hugo wrote, " Where the telescope ends, the microscope begins. Which of these has the grander view?" So you don't have to be a starship captain to "explore strange new worlds " or " go where no man has gone before." You only have to be a biologist with a microscope and an assortment of amber. Every once in a while a nugget will contain a surprise. Perhaps there will be a new species, genus, or even family of organisms previously unknown to science. It might be the first fossil record of an animal or plant, like the oldest bee or grass or even a strange chimera-like creature that is both mysterious and baffling. These discoveries are scientific jackpots that await the paleontologist like a winning ticket in the lottery, and they bring to mind one of our favorite stories, involving an amber dealer and an amber pendant worn to a party by an unsuspecting woman. Since any true amberphile is all too willing to look at any piece of amber to see what treasures lie within, it was only natural for him to examine the pendant. What he saw was amazing—two fleas. Did we mention that fossil fleas are extremely rare and here were two in one piece of amber? The pendant ended up in a museum! This

FIGURE 2. The preservation of extremely fragile objects, like this Burmese amber biting midge caught in a spider web, is one of the wonders of amber.

just goes to show that any piece of amber jewelry, including any bead in a necklace, can contain rare scientific treasures.

Can you believe that this warm, organic gem began as resin slowly oozing down the bark of an ancient tree millions of years ago? Amber is a fossilized resin, but the processes that turned it into amber are quite separate from those that preserved the insects embedded within. So amber is really a compound fossil repository containing not only evidence of the giant trees from which it came but also the creatures from that paleoenvironment.

After millions of years of being held deep in the earth's embrace or even immersed in the sea, amber from the Cretaceous has surfaced in various deposits around the world. All amber from that period seems to have been formed by an ancient group of conifers, the araucarians (fig. 3). These trees had a global distribution in the Cretaceous as seen by their fossil remains (fig. 4), although they are now limited to a few populations in the Southern Hemisphere.[344]

In some areas amber is dug from deep within the earth, but in southern Alberta, Canada, Cretaceous amber is strewn over a desolate prairie where wind and rain have scarred the land and released this treasure from imprisonment in coal and shale. When we traveled to those plains in western Canada with Ted Pike, then a doctoral student at the University of Calgary, we helped collect amber nuggets scattered among fragments of lichen-covered dinosaur bones. They ranged in size from teardrops to walnuts, and many were incrusted with a dark oxidized covering.

When you see them lying side by side, the contrast between amber and other fossils becomes apparent (fig. 5). Unlike entire insects trapped in amber, only the bones, teeth, claws, a few eggs, coprolites and footprints of dinosaurs are left for us to interpret. We only had to travel a few miles north to Dinosaur Provincial Park and the famous Tyrrell Museum of Paleontology to see one of the world's largest collections of Cretaceous vertebrate fossils. These were formed after their remains were inundated with mud or sand, infiltrated with waterborne minerals,

FIGURE 3. Photo of one of the largest living araucarians, a kauri tree, *Tane Mahuta*, located in the Waipoua Forest, New Zealand. The tree stands 169 ft (51.5 m) tall, has a girth of 45 ft (13.77 m), and is believed to be about 1,500 years old. The combination of height and girth indicates that it contains some 8,634 ft$^3$ (244.5 m$^3$) of timber.

**Cretaceous Distribution of Araucariaceae**

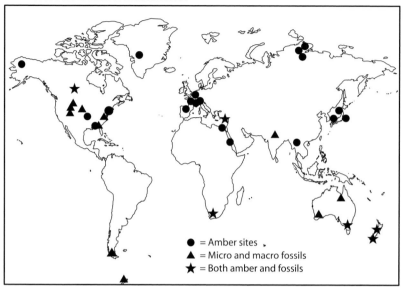

FIGURE 4. The global distribution of Cretaceous Araucariaceae based on amber, macro or micro plant fossils, or a combination of these. Fossils indicate that these coniferous resin-producing trees occurred worldwide. The isolated triangle at the bottom of the map represents an Antarctica araucarian fossil site.

and eventually turned into stone. Rarely, when the sediment covering a dinosaur was particularly fine, an impression revealed the texture of the skin.

Plant fossils have a different appeal than amber or bones. While amber has to be polished to reveal hidden treasures and dinosaur remains usually have to be meticulously excavated and transported to a lab for scrutiny, plants in all their glory can be instantly revealed at the collecting site—not that it still isn't hard work. We have spent hours scrambling up hot, slippery cliff faces covered with haphazardly positioned boulders. Despite the perspiration and the need to remain on the lookout for hidden rattlesnakes, the perfect rock sample could make the hunt worthwhile. When turned on the end and split, a sandstone slab would ideally display two plant impressions, one revealing the top and the other the bottom of a leaf surface. Some of these compression

FIGURE 5.  Different fossil types. *Top:* three Late Cretaceous dinosaur bone fragments; *middle:* leaf impressions of a Jurassic araucarian (*Agathis jurassica*) from Australia; *bottom left:* a Jurassic permineralized araucarian cone (*Araucaria mirabilis*) from Patagonia, Argentina; *bottom right:* pieces of Burmese amber.

fossils may occasionally retain areas of green pigment or even evidence of insect damage. What an elating experience to free a leaf from that stone tomb after millions of years.

Not just leaves, but petrified wood, trees, cones, and even pollen can tell of past forests. The Petrified Forest of the Chinle Formation in Arizona contains the mineralized stumps of araucarians and even small bits of amber. A trip there gave us a concept of the immense size of the resin-producing trees in the Mesozoic. We were impressed by massive logs up to 200 feet long and 10 feet in diameter. Their condition indicates that they had been transported in rivers for quite a distance before finally being buried and fossilized. These logs are incredibly beautiful—all their tissues were replaced by varieties of quartz, including amethyst and rose quartz. The dominant color, however, is that of jasper and ranges from reds to yellows and browns. Although these araucarian trees are impressive macrofossils, plant microfossils like pollen and spores, while not as dramatic, can often provide a greater amount of information about the types of plants present in the past.

This brings us back to one of the reasons we find fossils so fascinating: each one has a story to tell. From their anatomy we can glean information about lifestyles and habitats. From fossil assemblages, we can deduce facts about animal and plant interactions, the ecology and climate of those past ages, and taken all together, what life was like millions of years ago during a particular time in the Cretaceous. This is what we hope to reveal to you: a picture of life 145.5 to 65.5 million years ago when insects and dinosaurs competed for available food resources, fed on each other, and suffered from parasites and newly evolving diseases. First, we need to explain how physical and biological changes over the course of the Cretaceous would have affected how insects bugged dinosaurs.

# 2.

## The Cretaceous: A Time of Change

THE SUN rose and set over 29 billion times during the Cretaceous. Each succeeding dawn and nightfall saw the birth and death of billions of organisms, and in every passing millennium, species arose or became extinct. Dramatic physical and biological changes molded the evolution of insects, plants, and dinosaurs during that period in the planet's history. Differences in insect taxa are evident in the amber fossils found in Lebanon in the Early, Burma in the mid, and Canada in the Late Cretaceous. Other fossil deposits tell us that similar changes also occurred in dinosaurs and plants.

### Geographic and Climatic Changes

In the 25–30 million years separating each amber site, continents drifted hundreds of miles, mountains formed, sea levels changed, volcanoes erupted and died, climates shifted, and earthquakes fractured the land. All of these physical transformations in the earth's architecture were the consequences of what is known as plate tectonics.

Geologists have made great strides deciphering the mysteries of the planet's history with the study of plate tectonics. They have established that the earth's surface is composed of a complex of continental and oceanic plates that are being continuously pushed apart in some areas and forced together in others, so that they are constantly slipping and sliding over, under, or against each other. Even today the earth continues to shift and in the most active areas of plate movement, earthquakes and volcanoes are a daily reminder of its restlessness. Although this movement does not appear to have been uniform over geological time, and long periods of near-stasis have been interspersed with periods of rather intense activity, the end result was that the geography of the Early, mid and Late Cretaceous worlds was significantly different.

When we are presented with a map of 130 to 135 million years ago, the most obvious difference from one of today would be the very close association of continents situated in the Southern and Northern Hemispheres.[9] The landmasses joined in the supercontinent Gondwanaland (Africa, South America, Antarctica, Australia, and India) were beginning to separate from each other. Africa and South America were still contiguous and remained in contact with the other southern continents via land bridges and island chains (fig. 6). In the northern regions the continents forming the Laurasian complex (North America, Greenland, and Eurasia) were also still closely interconnected. The flora and fauna were free to migrate over the entire conjoined landmasses, thus potentially allowing for the distribution of a specific genus or species of insect, plant, or dinosaur to become quite extensive.

Moving forward to 97 to 105 million years ago, another map would show that as the continents separated, a rise in sea levels had placed the interior of Burma and its amber-forming forest close to the water's edge (fig. 7). While the rising sea level inundated the interior of North America with a vast epicontinental ocean and reduced much of Europe to large islands, it also further opened the Tethys Seaway between the northern and southern continents. Antarctica had almost reached the position it occupies today at the South Pole.

By the time the Canadian amber was being formed some 77–79 million years ago, western and eastern North America were still separated by that large interior sea (fig. 8). The continents had continued moving and now resided close to their present locations. Such an extended period of geological isolation would have resulted in endemic species occurring on the eastern and western sides of continental North America.

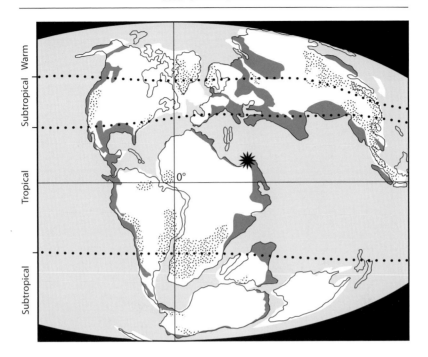

FIGURE 6. This map illustrates the putative positions of the continents when amber was being deposited in Lebanon (asterisk) during the Early Cretaceous. The probable extent of the land masses are in white, oceans are in light gray, epicontinental seas are in dark gray, and black lines delineate present-day continental perimeters. The dotted lines indicate theoretical tropical, subtropical, and warm climatic zones on land. The dotted areas indicate mountain ranges. (Modified from Smith et al.[9] for paleocoastlines and Vakhrameev[10] for climatic belts based on fossil plant distribution.) The horizontal and vertical lines intersect at 0° longitude and latitude.

We are left with the knowledge that geological changes over the course of the Cretaceous created an increasingly smaller terrestrial world as continents separated and sea levels rose. A team of scientists has estimated that from the beginning to near the end of the Cretaceous, there was a net loss of approximately 14% of non-marine areas on the globe so that at 80 million years ago, the world had the least amount of exposed land seen over the last 245 million years and an astonishing 28% less then it had 5 mya.[9]

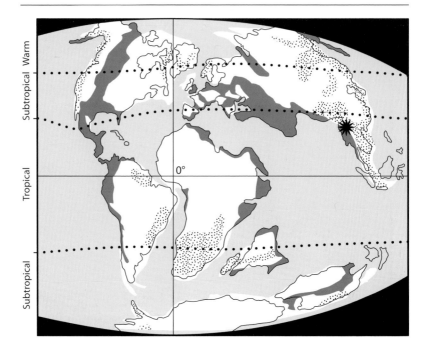

Figure 7. This map illustrates the putative positions of the continents when amber was being deposited in Burma (asterisk) during the mid-Cretaceous. The probable extent of the land masses are in white, oceans are in light gray, epicontinental seas are in dark gray, and black lines delineate present-day continental perimeters. The dotted lines indicate theoretical tropical, subtropical, and warm climatic zones on land. The dotted areas indicate mountain ranges. (Modified from Smith et al.[9] for paleocoast-lines and Vakhrameev[10] for climatic belts based on fossil plant distribution.) The horizontal and vertical lines intersect at 0° longitude and latitude.

Climate is another physical factor that influenced the lifestyles of past biota. Climatic changes have many causes, but landmass size and topography, along with wind patterns and ocean currents, contribute just as much as ozone layers and continental positions in relation to the solar energy gradient that extends from the equator to the poles. Since most of these can be correlated with changes brought about by plate tectonics, it follows that during the Cretaceous, the climate fluctuated somewhat. In general however, the Cretaceous was thought to have been con-

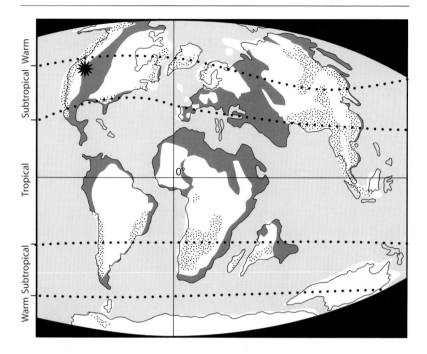

FIGURE 8. This map illustrates the putative positions of the continents when amber was being deposited in Canada (asterisk) during the Late Cretaceous. The probable extent of the land masses are in white, oceans are in light gray, epicontinental seas are in dark gray, and black lines delineate present-day continental perimeters. The dotted lines indicate theoretical tropical, subtropical, and warm climatic zones on land. The dotted areas indicate mountain ranges. (Modified from Smith et al.[9] for paleocoastlines and Vakhrameev[10] for climatic belts based on fossil plant distribution.) The horizontal and vertical lines intersect at 0° longitude and latitude.

siderably warmer than today, and it is unlikely that there were freezing temperatures for most of those 80 million years.

Various disciplines have methods to garner information about past climates, such as detecting the distribution of climatically sensitive mineral deposits in the earth's surface or examining the dispersion of past biota indicative of particular temperature preferences. Since our three amber deposits were found in coal strata, we can infer that their associated forests thrived in a moist to humid environment. We also know that the araucarian trees that produced the Cretaceous amber predominately prefer tropical to subtropical wet zones today. The shapes, edges, sizes, and thickness of fossil leaves often indicate whether a plant grew in a cold, hot, wet, or dry environment. Using these characteristics, the paleobotanist V. A. Vakhrameev created phytogeographic maps to delineate Cretaceous climatic belts[10] (figs. 6, 7, 8). Combining the available data leads us to portray the Lebanese and Burmese amber-forming habitats as lush hot and steamy tropical rainforests where the seasons were only marked by the amount of rain that fell each succeeding month. The Canadian amber forest flourished in verdant moist subtropical environs, and short mild winter days followed on the heels of long scorching summer days. All three forests grew near the sea, so they experienced a marine influence. Dense morning fog may have obscured the inhabitants, blanketed the trees, and deposited condensed beads of moisture on exposed surfaces. Periodic cyclones whipping across the vast tropical seas undoubtedly pounded the shores and wrecked havoc on the forests.

Alterations in global geography and climate over the course of the Cretaceous were accompanied by changes in the flora and fauna. The reasons for new plants and animals appearing while others died out are complex. The demise of each individual species had a unique set of causes and effects that were linked in an intricate web with the competition and survival of other organisms in their habitat or even the entire biome. Many of these biotic changes have been recorded in fossil deposits.

### Biotic Changes

As the predawn light began to infuse the Early Cretaceous skies over Lebanon, the creatures of the night sought shelter. Hunters and gatherers crept under leaves, sidled into crevices or caves, climbed toward the treetops, or scurried underground. For a brief moment, the forest was in repose. The coming blaze of sunrise would illuminate an almost alien plant world, quite different from what it is today.

The archaic plants that covered the land then were predominately gymnosperms such as conifers, ginkgos, and cycads (fig. 9). An important conifer from our perspective would have been the amber-forming araucarian, *Agathis levantensis*. Ancient ferns, liverworts, lycopods, equisetums, and mosses grew everywhere. Flowering plants, the angiosperms, were rare, diminutive, and inconspicuous, and mainly consisted of types with primitive archetypical flowers. The exotic seed ferns were reduced to almost relictual circumstances. Primitive shrubs like *Caytonia* with its cupule-covered seeds and related cycadeoids with barrel-like trunks dotted with reproductive organs lent a strange, mysterious atmosphere to the landscape.

The world was an unfamiliar evergreen place where wind-carried pollen and spores drifted through the forest ensuring successive generations. There were no showy displays of fragrant flowers to festoon open spaces with multicolored floral tapestries or dot the canopy with brilliant splashes of color. The few plants fossils that have been found at the amber site certainly do not represent the diversity of flora that must have been present.[11–13] There in Lebanon and throughout the globe were the last survivors of plant communities, the likes of which would disappear from the face of the earth by the mid-Cretaceous.

The composition of the flora of the entire planet was about to undergo a transformation as the angiosperms began to blossom and initiate the meteoric growth in diversity that would culminate in their dominance of the plant world (fig. 9). Already by the mid-Cretaceous, there were significant differences. Obvious changes in the 30-million-year span were that new genera and species of plants had evolved. Emerging habitats were replete with more modern conifers. In the Burmese amber-forming araucarian forests, aromatic cedars touched slender pines, lofty sequoias grew next to elegant dawn redwoods, and expansive cypresses made contact with compact junipers.[14–20] Ferns (fig. 10) and club mosses still dominated the undergrowth.

Those ascending stars, the angiosperms, were now aggressively competing for space everywhere, and their increasing

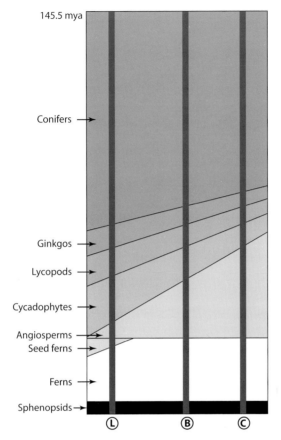

145.5 mya

Conifers →

Ginkgos →

Lycopods →

Cycadophytes →

Angiosperms →
Seed ferns →

Ferns →

Sphenopsids →

Ⓛ            Ⓑ            Ⓒ

FIGURE 9. The composition of global flora changed considerably over the 80-million-year span of the Cretaceous. Most significant was the radiation of the angiosperms. This was accompanied by a decline in diversity of the other plant groups. The letters L, B, and C indicate the approximate times that the amber forests were flourishing at the Lebanese, Burmese, and Canadian sites, and illustrate in a very general way the proportions of plant types that could have occurred at those locations.

diversity appeared to be accompanied by a decline in diversity of other plants.[18,348] Infrequent clusters of delicate pale flowers peeked out from verdant growth and palm nuts accumulated with conifer cones on the forest floor. At least two grasses (fig. 11) with affinities to present-day bamboos prospered,[17] possibly even forming ever-expanding thickets that threatened to overgrow the vast expanses of ferns and horsetails. Club and bracket fungi decorated decaying trunks while puffball and pinwheel mushrooms flourished on the forest floor[91,345] (fig. 12, color plates 12A, 12D).

By the time the Canadian amber was being formed, the forests, shrub lands, and meadows more closely resembled those of the present. Our knowledge of the plants growing in Alberta during

FIGURE 10. Tip of a fern leaf in Burmese amber.

the Campanian Stage of the Late Cretaceous is actually quite extensive due to the concerted efforts of many paleontologists working over decades.[21-27] In the forests, podocarps probably now mingled with the resin-producing kauri (araucarian) trees. The numbers of angiosperms increased significantly, and almost certainly some of the plants growing in the northern latitudes were now deciduous. The size and habits of those ancient flowering plants and how they may have contributed to the ecosystem is open to debate. Had some developed into trees by this time or were they only ground cover or low shrubs? As Dennis Braman and Eva Koppelhus remarked about Late Cretaceous fossil angiosperms in Canada, "If they had tree-like habits, one would expect to find large pieces of fossilized wood. However, angiosperm wood is presently unknown".[21] So we can only guess how they contributed to the strata in these forests.

There are many factors that determine where a plant will grow, including temperature, water availability due to precipitation and evaporation, soil type, drainage, light, latitude, terrain, and elevation. In any of the amber sites these parameters changed with time, affecting the plants that were able to flourish there in succeeding millennia. While the distribution and relative importance of the various plant types throughout the Cretaceous gradually changed with some families, genera, and species

FIGURE 11.
Spikelet of the
primitive grass
*Programinis burmi-
tis* in Burmese
amber.[17]

FIGURE 12. A group of puffballs in Burmese amber.

dying out and others converting into more modern forms, the animals that utilized them for food and shelter were also undergoing transformations.

We know that dinosaur species came and went. In fact, in the 25 to 30 million years separating each amber deposit, the fauna would have been replaced by new species and genera a multiplicity of times. Peter Dodson believes that individual dinosaur species lasted only between 1 and 2 million years, while their mean generic longevity averaged about 7.7 million years.[28] This suggests that the entire dinosaur fauna may have turned over with every geological stage, or approximately ten times in the Cretaceous.[29] With a high turnover rate, relatively few fossil sites, and the entire global distribution of dinosaurs over a vast time period to be considered, only some general patterns of the changes can be accessed.[30–32] Unfortunately, while the presence of a fossil tells us what lineage existed in a particular time and place, the absence of other organisms does not mean they did not exist during that period.

In the Cretaceous, a gradual shift in the overall size of plant-eating dinosaurs from the gigantic multiton behemoths once dominant in the Jurassic to relatively smaller and faster herbivores occurred in the Northern Hemisphere (fig. 13). The ornithopods in particular proliferated and spread. Perhaps the best known of these in the Early Cretaceous were the iguanodons, which were quite common in the northern hemisphere, while closely related ornithopods roamed the southern continents. Ponderous ankylosaurs, embellished with heavy dorsal bony plates decorated with an array of horny studs and spikes and armed with a clubbed tail, became prominent components of early Cretaceous fauna in the areas once comprising Laurasia, but were rare in Africa. Stegosaurs, those plated dinosaurs with ridiculously tiny heads and two rows of spines jutting from their backs, became extinct or scarce in some regions. In those habitats where they managed to survive, they did not appear to maintain a very significant presence. The small psittacosaurs, primitive ceratopsians with parrot-like beaks, became a successful group

FIGURE 13. Cretaceous herbivorous dinosaurs, showing their relative sizes. Some of the herbivorous/omnivorous dinosaur morphotypes that could have lived in or near one or more of the amber sites are shown here. By far the largest land animals known were the sauropods. Examples of these were brachiosaurids (1) and titanosaurids (2), as well as diplodocids and cetiosaurids. Other massive four-legged herbivores, but having horny "beaks," included stegosaurids (3), ankylosaurids (4), nodosaurids (5), and ceratopsids (6). A small quadruped group of dinosaurs were the protoceratopsids (7). Walking on two legs or sometime four were the ornithopods, which ranged in size from very large to comparatively small and also possessed beak-like structures. These included iguanodontids (8), hadrosaurids (9), dryosaurids (10), and thescelosaurids (11), intermediate-size ornithopods. Psittacosaurids (12) were small ceratopsids that probably walked on hind legs. Pachycephalosaurids (13) also walked on their hind legs and varied in size from small to intermediate.

in Asia. Herds of sauropods, the largest of the quadruped herbivores, with incongruously small heads perched on long necks, bodies the size of a house, and legs like telephone poles, continued to lumber with earth-shaking footsteps across the horizons of South America. These behemoths, 80 to 100 feet long, are estimated to have weighed up to 50 tons. They were seen, however, with ever-decreasing frequency in other parts of the globe.

By the Late Cretaceous new groups of dinosaurs begin to make their appearance for the first time in the Northern Hemisphere. Small pachycephalosaurs, many with prominent thick-roofed skulls crowned with rows of bony knobs, and a variety of ceratopsians with horns, frills, and oversized heads now joined herds of armored ankylosaurs and bipedal/quadrupedal hadrosaurs that could reach 40 feet long and weigh over 4 tons. Surprisingly, it appears that the southern continents retained mostly Early Cretaceous dinosaur types.

As the herbivores moved from large and slow to smaller and faster, and some became armed with defensive horns, the carnivores that fed on them changed also (fig. 14). The first part of the Cretaceous saw a burst in theropod diversity. These fierce, two-legged meat-eating machines became larger and if possible more formidable. Deinonychosaurs armed with capacious sickle-shaped claws appeared in the Northern Hemisphere. Members of this group included the dromaeosaurs and troodons that possessed relatively large brains and eyes, and were equipped with retractable second toes ending in a curved claw that could be flicked forward to slash their victims. The therizinosaurs, also in this group but apparently confined to Asia, were not only large but each hand terminated in three terrifying scythe-like claws.

By the late Cretaceous, the largest of all terrestrial carnivores now ruled the earth—the tyrannosaurs. Every child can easily envision *T. rex* with its characteristic immense jaws, 10-inch curved fangs, 4- or 5-foot-long head, and 45-foot-long body. Their overwhelming stature was somewhat offset by curiously short arms ending in tiny two-fingered hands. Other more obscure but equally foreboding theropods were the carcharodon-

FIGURE 14. Cretaceous theropods, showing their relative sizes. This figure illustrates some of the theropod morphotypes that could have frequented one or more of the amber forest sites. Theropods were a group of bipedal dinosaurs that included ferocious predators but also contained omnivores, insectivores, and perhaps even herbivores. They ranged in size from no bigger than a chicken to forty feet long. Some were small and armed with sharp teeth and curved claws, such as the coelurids (1) and deinonychosaurs such as dromaeosaurids (2) and troodontids (3). These were considered swift hunters that traveled in packs. Other small- to intermediate-size theropods but with toothless beak-like structures were oviraptors (4), garudimimids (5), and ornithomimids (6). All of these resembled the ratite birds of today. Dominating the land were the large theropods. Spinosaurids (7) were unique in that they had a skin sail supported by tall spines along their back. Abelisaurids (8) had deep snouts and flat horns over their eyes. Therizinosaurids (9) had long forelegs, and based on their teeth type and beak, were probably herbivores. The best known of all theropods, the tyrannosaurids (10) epitomize the flesh-eating top predator of the Cretaceous.

tosaurs and the gigantosaurs. Although sometimes considered scavengers, all of these megacarnivores may have had to compete for food with roaming packs of dromaeosaurs or troodons. Two other emerging groups of theropods, however, may have fed on considerably different food. The ornithomimids and oviraptorosaurs eschewed rows of pointed fangs for prominent toothless beaks that nevertheless retained considerable shearing power. The unique, somewhat relictual dinosaur fauna of the Southern Hemisphere included the abelisaurs, large horned theropods. While large thundering herds of heavy herbivores and huge cunning carnivores are most frequently fossilized, small dinosaurs ranging from the size of chickens to large dogs also abounded, and their importance shouldn't be overlooked.

### Insects

There are some 762 families of insects today.[33] If you multiply that number by a factor of four, you will probably have an idea of insect diversity in the Cretaceous, since considering how few Cretaceous insect fossil sites there are, already some 490 families are known from that period (appendix A).

Changes in insect diversity can be demonstrated by comparing fossils from the Early, mid, and Late Cretaceous. Many Early Cretaceous insects retained archaic characters, physical attributes unique to each species, from their Jurassic forebearers, and this is reflected in representatives from Lebanese amber. These include primitive weevils[34] that developed in conifer cones (fig. 15), bristly walking sticks, giant katydid-like insects, and stemboring sawflies, all of which fed on gymnosperm leaves or pollen.[35] They shared their world with spiny cicadas feeding on ginkgos and primitive moths whose caterpillars consumed moss.[36] Small thrips[37] with mouthparts modified for piercing and scraping devoured pollen and even defoliated some of the araucarians, while long-beaked aphids[38] sucked the juices of conifers, horsetails, ferns, and cycads.

It was just about this time when some insects developed a

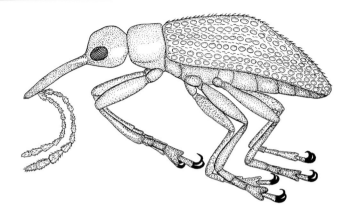

Figure 15. Larvae of this small weevil could have developed in the male cones of *Agathis levantensis*, the Lebanese amber-producing tree.[34]

taste for vertebrate blood. By the Early Cretaceous, horseflies, blackflies, mosquitoes,[35] biting midges, sand flies, and a host of lesser-known hovering insects were all on the lookout for the fresh blood of everything from small frogs to gigantic dinosaurs.[13]

The diversification of the flowering plants in the mid-Cretaceous spelled extinction for many archaic insects. In fact, there were more hexapod extirpations at this time than even at the K/T boundary.[35] The mid-Cretaceous was a time when older, established insects competed with modern ones for the same habitats. Some of the primitive types were inconspicuous beetles so small that hungry lizards patrolling the bark would have probably ignored them (fig. 16). One such beetle fed on mites and possessed an extra segment that provided its front legs with an added degree of flexibility to help snatch prey.[39] Sharing the same habitat was a strange beetle with cuticular modifications on the head and body rivaling the frills, horns, and plates of any dinosaur. Not only did this insect have shield-like projections that protected bulging eyes, but also a fin-like protrusion on the top of the head plus a series of frills that ran the length of the back.[347] A third beetle was so archaic it had to be described in a new family.[40] With reduced mouthparts, this insect could only have fed on fungi, moss, or algae. Other beetles used chemicals for defense.[349]

FIGURE 16. Three fascinating beetles from Burmese amber. In the upper right is an extinct subfamily of ant-like stone beetles (Scydmaenidae).[39] The unique character of this little mite predator is the extra segment in its front legs. The upper left shows an extinct silvanid beetle with head and body protrusions that could rival most dinosaurs. Its large bulging eyes were protected by triangular projections.[347] In the center, hiding in a hollow and partly covered by a moss leaf, is the only known member of the extinct flightless beetle family Haplochelidae.[40]

Larger insects, which had evolved millions of years earlier and were near the end of their reign, included creatures that looked like a cross between a grasshopper and cricket (color plate 8A). These sturdy elcanids, with protruding eyes and long antennae, probably fed on gymnosperms. They would have interacted directly with dinosaurs by providing a food source, as well as competing with them for dwindling plant resources. However, their tiny cousins, pygmy mole crickets and pygmy grasshoppers, would have attracted the attention of hungry geckos and amphibians (color Plate 7B).

Evidence of more modern insect lineages at the mid-Cretaceous Burmese amber site included a weevil with elbowed antennae similar to those that feed on flowering plants today (color plate 5E).[41] Insect damage to angiosperm flowers in Burmese amber shows that herbivores had already adjusted to a diet of flowering plants[18,19] (color plate 14A). Perhaps the injury was from a small, slender thrips or the larvae of a gall midge (color plates 3A, 3C). Certainly there would have been a horde of insects, including planthoppers and aphids, feeding on primitive grasses, and they competed with dinosaurs for that delicacy[17,42]. Some of those aphids survived very well with only two wings instead of the four found on plant lice today.[43] These sucking insects could have been carrying plant viruses then. A surprise find was a small bee that was certainly pollinating some of the angiosperms at the site[44] (color plate 14B).

Aside from the typical bloodsuckers, ticks also occurred at the Burmese amber site[45] (color plate 11E). It is a challenge to imagine what other arthropods with piercing mouthparts were feeding on in that amber forest and whether they played an important role in the transmission of disease-causing pathogens[46,47](color plate 10). An unexpected surprise was two-winged scorpion flys with beaks containing dagger-like mandibles with serrated edges (color plates 11A, 11B). We can only guess how many other types of biting arthropods were present in those Cretaceous forests.

By the Late Cretaceous, more of the insects resembled those

found today, including a palm-feeding beetle at the Canadian amber site[26] (color plate 5A). There must also have been a considerable number of new herbivorous insects related to those that feed currently on maples and sycamores, alders and elms, or lilies and sedges, all plants present at that time (color plate 12B). Countless leaves were probably covered with the sweet deposits and shed skins of plant lice. The high number of aphids in Canadian amber, amounting to about 40% of the total animal fossils, indicates that they were one of the dominant groups of insect herbivores[49] (color plate 2C). By then some females had developed the ability to eject squirming young directly on the plant and forego an egg stage, an advantage their ancestors lacked. Since extant aphids carry approximately 50% of insect-transmitted plant viruses, perhaps they were responsible for the rapid turnover of Late Cretaceous plant lineages. A mosquito in Canadian amber indicates that these bloodsuckers were thick at that particular location, possibly breeding in salt marshes near the amber site[191] (color plate 11D), while the mouthparts of biting midges suggests that they fed on dinosaurs.[51]

Insect lineages came and went throughout the Cretaceous, and most of them had either a direct or indirect association with dinosaurs. Some bugs simply provided a food source, while others ate away at the dinosaur's plants or spread blights or other plant diseases. Scavengers recycled dinosaur waste and cadavers, and the bloodsuckers, at the top of the food chain, not only feasted on dinosaurs but also introduced pathogens into their bodies. What follows is an intriguing story about the struggle between plants, insects, pathogens, and dinosaurs that took place millions of years ago.

# 3.

---

## Herbivory

*The sun beat down on the canopy of the freshly washed forest, causing steam to rise like puffs of smoke from the wet vegetation. Emerging through this leafy roof were the foliage crowns of the dominant trees, the kauri trees (araucarians). Massive trees with bases up to 40 feet in diameter, they reached skyward 120 or even 200 feet. Their flattened, elliptical leaves fluttered in the sunshine, displaying tapered ends, prominent longitudinal veins, and insect damage. A kauri cone infested with small caterpillars released its hold high in the branches, fell spiraling down through the canopy, dislodging leaves, ricocheting off the trunk, and coming to rest in the copious pile of duff built up around the base of the centuries-old tree.*

*The noise alerted a group of juvenile pachycephalosaurs that were foraging in the dimly lit recesses on the forest floor. One moved to the area where the cone had landed to poke and paw through the debris. He picked at the infested cone and was rewarded with several juicy larvae to eat. The pachycephalosaurs were dwarfed by the surrounding trees and lush greenery. Elegant conifers towered above the forest's floor like cathedral spires looming over a cityscape, while other species of archaic gymnosperms formed the dense canopy. Their narrow, pointed, and frequently prickly foliage was quite different from the exotic vein-ribbed leaf fans of the surrounding ginkgos, which also pushed upwards to share the light illuminating the canopy layer. Their mingling leaves carved by insect bites into strange sculptures fluttered in the wind and created a symphony of green hues delicately shaded with occasional glints of russets, browns, and yellows.*

*The young bipedal dinosaurs moved from the dim recesses of gloom-filled aisles into sunlit pools of open spaces created by fallen giants and along the bank of a dissecting stream, through an understory flora of evergreen trees mingled with wooly tree ferns and coarse palms. They feasted on ferns, which grew everywhere in a profusion of sizes and shapes with leaves as delicate as lace or as stiff as bristles. Some, like the rugged climbing ferns that festooned the limbs of the mighty trees or sprouted from trunks, were too high to reach. But the smaller types crowded along the riverbank or flourishing in meadows were a delicacy. Male and female cycads, looking like stubby palms, vied for space with large ferns, squat conifers, and the primitive relatives of bushy magnolias with small fleshy fruits, and all offered up some tidbits to eat. Any habitable space on the forest floor was covered in a luxuriant undergrowth of ferns, equisetums, small herbaceous plants, and hopeful seedlings providing habitats and food for a multitude of small vertebrates and invertebrates. Moist undulating pockets of shade carpeted with a fabric woven of delicate bracken ferns, pale ghostly lichens, and dark velvety mosses were home to mites, springtails, and spiders.*

*Lianas and epiphytes festooned every layer in the forest. These effectively transformed the canopy into a roof garden where they clung to the moss-covered branches and peeked through the leafy cover. The trunk and limbs of trees were draped with swaths of delicate mosses, sheets of liverworts, and veils of filmy ferns. Ropy lianas stretched across and around trunks and limbs, binding together every level in the forest and serving as highways for cockroaches and crickets. They intertwined with masses of tangled roots to form a meshwork on the forest floor, and twisted and looped out over the top layers of the canopy. Epiphytes perched in every stratum, weighing down limbs, creating dense luxuriant growth over every available surface, and providing cover for frogs and birds. Even sedges grew in the forks of the trees, while feathery bamboo-like grasses flourished in the open areas. At ground level, a litter of discarded leaves, cones, twigs, logs, bark, and primitive fruits made an ideal home for saprophytic fungi of all descrip-*

tions, as well as centipedes, millipedes, and scorpions. And deep down in the loamy soil, the conifers of the forest spread stabilizing roots that coexisted with the rhizoids of symbiotic fungi and provided food for nematodes, beetles, and isopods.

This verdant jungle was teeming with hidden life. The pachycephalosaurs were not the only animals feeding there, only one of the more visible forms. Thousands of species of insects along with a sizable contingent of other arthropods crawled over every tree and bush, munching and crunching their way through a lavish feast. Beetles burrowed secretively under bark or browsed on pollen. Caterpillars grazed concealed on the underside of leaves while planthoppers covertly sucked plant juices. Camouflaged crickets sang and the calls of unseen strident grasshoppers permeated the air. Herbivorous insects subsisted on practically every plant surface at every level throughout the forest. Foliage was devastated as leaf beetles skeletonized, small moth larvae mined, crickets ate holes, and caterpillars gouged out the leaves. Lizards, frogs, and small mammals ran over the branches and crossed the lianas in search of prey while other warm-blooded vertebrates combed the plants for nutritious seeds and edible leaves.

On the forest floor, the group of small pachycephalosaurs continued to search for mushrooms, fallen fruits, and open seed cones to devour. One broke open a rotten log and the exposed fat white beetle grubs were gobbled down by these opportunistic omnivores. Suddenly cautious, they paused and looked in unison towards the cracking sounds now emanating from ahead. Like silent ghosts, a large herd of forest-dwelling ceratopsians had moved unnoticed through the undergrowth and were now using their massive heads to push over succulent saplings and partake of their leafage. Disturbed by the hubbub, a solitary male ankylosaur that had been feeding quietly on a patch of equisetum crashed angrily through the tangled underbrush in a rush to get away. A flock of ostrich-like ornithomimids striding along a pathway forged by the daily movements of countless large herbivores stepped off into the undergrowth and used their beaks to strip *leaves and insects from the surrounding shrubs and low-growing herbaceous plants. Their movements had been shadowed by several juvenile troodons forced from the pack by a new leader. They hadn't eaten in almost a week and the ornithomimids looked like an easy meal. Creeping closer, a stronger more aggressive troodon, emboldened by hunger, leapt out to snatch the closest victim. The two dinosaurs leapt into the air, kicking out at each other with their hind legs. But the ornithomimid was larger and more experienced than the attacker, and his longer legs armed with toe claws carved a bloody swath down the chest of the inexperienced carnivore, forcing him to retreat along with his companions. Later the immature theropods would settle for a meal of insects and a few mouthfuls each of a snake they had fortuitously encountered. None of the herbivores went hungry because this lush tropical rainforest provided a cornucopia of plants to feast upon.*

*Signs of the feeding activities of insects were everywhere. Horsetails were turning pale and shriveling up because of the beetle larvae in their stems. Ginkgos had been denuded of florets and leaves due to an infestation of moth and sawfly caterpillars. Cycads and cycadophytes showed insect feeding activity with gaping holes in their leaves, and their brown-stained stems were filled with succulent moth and beetle larvae. The fronds of stunted tree ferns had died back as weevil larvae surreptitiously tunneled through their tissues. Smaller ferns had been stealthily defoliated by leaf-feeding sawflies and moths. Some leaves of the giant kauri trees were twisted and discolored from numerous leaf mines caused by small moth larvae, while huge holes in the leaves testified to the presence of giant weevils bristling with clusters of erect hairs that crept furtively along the branches. Fallen limbs exposed honeycombed wood, which served as brood chambers for bark beetles and weevils. In fact, entire areas of the forest contained dying kauri trees. The needles of* Metasequoia *trees had turned yellow due to the feeding of multitudes of scale insects and aphids even while stout leaf beetles and long-horned grasshoppers fed on the foliage. Patches of needles on various conifers had been covered and bound together with webbing made by gregarious sawfly larvae, and the activities of voracious sawfly caterpillars, walking sticks, and grasshoppers had partially denuded large branches.*

*Flowering plants had not escaped notice by the insects. Notches along the edges of many small flowers on both herbs and shrubs indicated that chewing insects had added them to their diet. Leaves on some had been "sewn" together with threads secreted from the silk glands of small caterpillars, creating a safe refuge where they could feed unmolested. Rolled up palm leaves with their inner epidermis scraped away and older leaves with huge incisions extending from their edges to the midvein were evidence of the activities of different stages of white-lined green caterpillars. Weakened by large white weevil larvae chomping through their tissues, toppled palm stems lay scattered on the forest floor. In a nearby pond, many water-lily leaves had been consumed by populations of leaf beetles. Ancient bamboos climbing up the trunks of the araucarians showed extensive leaf damage from large brown scarab beetles, and their roots suffered from attacks of small caterpillars. The hidden hordes of insects were gradually consuming their share of the forest's abundant flora.*

Green plants are the most important organisms in any terrestrial ecosystem and have been since they first evolved. In terms of diversity, they are second only to insects and represent about one quarter of the total known species. They capture the energy from the sun and convert it into the plant tissues that sustain all animals. Herbivores transfer that energy directly into their tissues when eating plants, and carnivores indirectly obtain the energy when feeding on herbivores. For this reason plants comprise the very basis of the food chain.

There are many factors that determine where a particular insect or dinosaur lived, but food supply is the most important. The quantity of plants and their food quality determines how

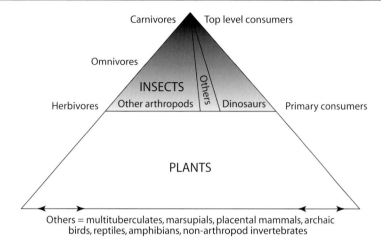

Others = multituberculates, marsupials, placental mammals, archaic birds, reptiles, amphibians, non-arthropod invertebrates

FIGURE 17. An ecological pyramid can be created based on inferred numbers, biomass, or energy utilization in the Cretaceous. The actual size and proportions of the pyramid would be unknown, but insects and other invertebrates definitely dominated any ecosystems then as they do now. Dinosaurs would have represented a much smaller percentage and would also have had competition from all the other vertebrates. Ecological pyramids are not static, and contract and expand continuously due to many biotic and abiotic factors. Different pyramids can be created for large and small, more specialized habitats.

many and what types of herbivores and ultimately carnivores reside where. Biologists have constructed ecological pyramids to illustrate this principle (fig. 17). There are several variations: the numbers pyramid (how many organisms), the biomass pyramid (total weight of all organisms present), and the energy pyramid (how much energy is produced, used, and stored in organisms). The pyramids are divided into several levels. The basal and largest contains the producers, the plants. The succeeding ones are devoted to the consumers, the animals. The consumers are further subdivided into types or orders. The first type or primary consumers are herbivores (plant predators), the second type are carnivores (animal predators) that feed on herbivores, and the third type are carnivores that feed on other carnivores, and so on until you reach a top predator. Omnivores fit into a ubiquitous category that extends across all the consumer levels.

Most people don't realize that insects dominate the typical consumer levels of any ecological pyramid, whether it denotes numbers, biomass, or energy, and they have probably occupied that position since they first walked the earth. They represent up to 60% of the herbivorous consumers in most ecosystems. In addition, insects and other small arthropods consume significantly greater amounts of plant tissues than all the vertebrates in any habitat (except grasslands) studied, and this prevalence has existed since the earliest of times.[53,55] This means that the remaining 40% of the consumer portion in a pyramid of 100 mya would have been subdivided between the dinosaurs and all other Cretaceous animals such as mammals, reptiles, amphibians, birds, and invertebrates (fig. 17). The exact proportions that each of these would share is not answerable. But however large the piece of the pyramid that dinosaurs would get, we know that it would be significantly less than that of the insects!

An ecological pyramid should be viewed as a continuously expanding and contracting entity. Its configuration changes from one ecosystem to the next. A tropical rainforest with the greatest diversity of species anywhere on earth would represent the consummate pyramid. This type of environment certainly covered more of the globe in the Cretaceous. Half of the known species in our world reside in tropical rainforests, which presently cover some 6% of the earth's land surface.[66] Insect species there are estimated to number between 5 and 10 million. Since a larger portion of the earth was tropical rainforest habitat in the Cretaceous, it follows that greater insect diversity occurred then. The opposite end of the pyramid spectra would be one with sparse vegetation, such as a desert or tundra. There is in fact a 75-fold difference in plant biomass between a rain forest and tundra, and a 20-fold difference between a rain forest and a marsh.

Smaller, specialized pyramids can be applied to different levels in the biome, from communities to niches. Within each ecological pyramid there are one or more keystone species whose presence is crucial to its stability. All other species within an ecosystem have different levels of importance. The interrelationships can change overnight and a pyramid can collapse at any time. Any collapse would then be followed by a recovery period, and the pyramid would expand until equilibrium between plants and animals was once again obtained. But recovery may not result in the same populations of flora and fauna. Contraction and expansion of an ecological pyramid is normally cyclical or seasonal, but can be catastrophic due to fire, drought, floods, overgrazing by transients, and so on.

Within each ecosystem there are a variety of habitats with specialized niches. The animals compete within these for food, space, and shelter. In a forest these habitats are distributed throughout vertical strata. The rain forest has the open air space above, and inside the forest, the emergent layer, the canopy, the understory, the shrub layer, the herbaceous layer, the litter zone, and the soil are usable habitats. To reconstruct a scenario from ancient landscapes, the paleoecologist tries to fit the fauna into these levels (fig. 18). The air above the forest would have been home to gliding pterosaurs and archaic birds (feathers have floated down through the trees to be entombed in amber at all three amber deposits: color plates 15B, 15C). The birds pursued insects soaring above the crowns of the trees in the hot humid air or alighted on the branches to snatch insects feeding there, while plant-eating birds fed on seeds, pollen, and fruits. Small pterosaurs chased dragonflies and large ones may have swooped down and yanked unsuspecting mammals from the treetops.

Scampering throughout all levels of the forest and leaving hairs behind in the resin were small mammals, multituberculates, and marsupials that had developed a scansorial lifestyle. Most of these were herbivores, omnivores, or insectivores. Joining them, moving along the lianas or among the leaves and even hiding in epiphytes, would have been a myriad of other animals such as tree frogs, lizards, snakes, geckos, hundreds of thousands of insects, and innumerable other small arthropods (color plate 15A). Each species of animal would find its own particular niche, a physical, environmental and biological space suited to their specific needs. At the ground level for example, there

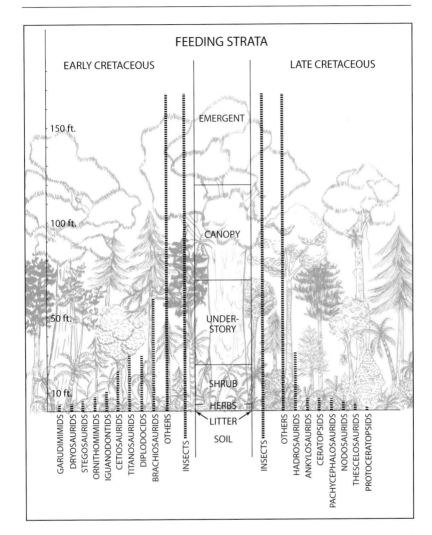

FIGURE 18. Forests in the Cretaceous would be divided into strata where animals and plants lived and competed. Based on extant kauri forests, the layers would be: soil, litter, herbaceous (up to 3 ft), shrub (3–25 ft), understory (25–70 ft), canopy (70–120 ft.), emergent (above 120 ft), and free airspace. The structure of the forest differed in the Early to mid-Cretaceous from that of the Late Cretaceous due to changes in plant community composition and the radiation of the angiosperms. Also, the dinosaurs were smaller and fed at lower strata in the Late Cretaceous as shown by the columns indicating their estimated feeding levels. Insects exploited all strata from the soil to the air, but competition with dinosaurs for food was confined to the lower levels. The large trees with open foliage represent the amber-producing araucarian trees.

would be insects, mammals, and even dinosaurs adapted to a fossorial life style. These would root around in the leaf litter, scratching in the soil or digging burrows. Mites and millipedes, springtails and crickets, annelids and nematodes, snails and scorpions, as well as cockroaches would have been creeping or slithering through the leaves, eating and being eaten. Insects moved and fed in every layer of the forest. They devoured the foliage, flowers, fruits, seeds, wood, pollen, nectar, and even resins. Perhaps three hundred or more insects fed on a single tree, but many of these probably included only three or fewer plant species in their diet.

Fitting dinosaurs into the forest takes a little imagination. We assume that many of the dinosaurs living in forests were restricted to feeding in the lower strata. No one currently seems to believe that non-avian dinosaurs resided in or even climbed trees. Size might have limited the extent to which extremely large dinosaurs entered the forest, and these may have been relegated to feeding along the fringes or in more open woodlands. Certainly small to intermediate bipedal and quadrupedal dinosaurs moved quietly through the dark, dense humid rain forests searching for food.

Animals that did not fly or climb had to feed within a vertical range dictated by their height or find some other method to extend that range. It has been suggested that some quadruped dinosaurs adapted a tripedal position and balanced on their tails and hind legs to achieve a feeding stance that extended their height.[52] Sauropods, already outfitted with extremely long necks, and stout stegosaurs were suggested as likely candidates (fig. 19). The feeding strata of dinosaurs therefore had a wide potential (fig. 18); however, we will never know the true extent. Most earthbound animals today spend the majority of their time feeding at eye level or below, with some very few exceptions. Overall, the megaherbivores of the Late Cretaceous fed at a lower level than those of the Early Cretaceous, at least in the Northern Hemisphere. An Early Cretaceous sauropod could compete with insects feeding well into the understory. A Late Cretaceous cer-

FIGURE 19. An Early Cretaceous landscape. In the foreground two juvenile coelurids search for insects among the ferns, a dragonfly rests on ginkgo leaves, and an insect scurries up the trunk of an *Agathis* tree. In the middle, an iguanodontid feeds on cycads under the watchful eye of a spinosaurid hidden among some tree ferns. In the background, two diplodocids feed on *Agathis* foliage at the edge of an amber forest. The trunk of the resinous *Agathis* tree shows its cones and leaves.

FIGURE 20. Hind leg and portion of tail of a gecko in Burmese amber.

atopsian would only look for food in the low shrub to ground level, and the protoceratopsians coexisted with insects in the herbaceous layer of the ecosystem. But in araucarian forests, all the space above the understory was left to the insects, other arthropods, and small vertebrates (fig. 20).

Herbivores are the most numerous and diverse animals in any ecosystem. So how did phytophagous insects and dinosaurs compete? Probably in much the same way that insects are known to compete with mammals (including ourselves). Since insects feed on plants at all levels, we surmise that whatever the dinosaurs ate, insects were also there dining on it. We know that insect herbivores have greater effects on both plant growth and reproduction than mammals,[53] and there may be up to ten times more invertebrate plant eaters present in an ecosystem than mammals.[54] These same observations undoubtedly applied to ancient ecosystems; the difference in impact of megaherbivores vs. insects on the plants lay in the greater size of the dinosaurs. However, remember that pound for pound, the insects in most ecosystems would have outweighed the dinosaurs.

Thus, one megaherbivore may have eaten as much as 100,000 insects or more, but the general impact of millions upon millions of insects would have been more gradual and widespread. Large herbivores with their greater body size tend to eat a wider range of plants, take larger bites, and move over a greater area, trampling, defecating, and urinating copious amounts. This cuts a swath of defoliation and destruction across a narrow section of the terrain in a short amount of time. Insects with their small size but incredible numbers may eat as much or more, but they are normally more evenly distributed throughout the ecosystem. Because they are small, insects can selectively feed on seeds and roots that have a cumulative impact on plant viability. Insects also carry plant pathogens that can devastate entire plant populations. So while small and inconspicuous, insects probably were serious competitors with dinosaurs, not only eating more of the plants but also affecting overall plant health and distribution.

# 4.

## Dinosaurs Competing with Insects

FROM THE HABITS of present day herbivorous insects, we can infer how they would have competed with dinosaurs. We know that nemonychid weevils, like the one found in Lebanese amber, feed on pollen in the male cones of kauri trees and presume that they had similar habits in the Early Cretaceous. And it is quite likely that this source of protein was sought after by dinosaurs, just as birds and lizards feed on pollen today.[56] The interfaces between insects and dinosaurs regarding conifer cones represent just one type of antagonism that would have occurred between these groups. Dinosaurs and insects competed in the understory and lower levels of the forest and were the most serious threats to saplings and other new plants. Insects also fed in the upper canopy and emergent layers of the forest where their damage affected the health, longevity, and reproductive capabilities of the trees.

While few insects are known to ingest horsetails, certainly an entire guild feasted on the diverse assemblage of these coarse plants in the Cretaceous. Some insects, such as stem-inhabiting beetles and leaf-feeding caterpillars, still utilize these primitive plants, and if their ancestors had similar habits, herbivorous dinosaurs could have suffered.

Throughout the Cretaceous, ginkgos, or maidenhair, trees were diverse and widespread, and the periodic appearance of their pungent, cherry-like fruits certainly attracted both insects and dinosaurs. Protoceratopsians and pachycephalosaurs probably never missed the opportunity to consume these luscious fruits, but they must have competed with a succession of beetles and flies.

Dinosaurs that ate the leaves, peg-like stems, and male flowers of ginkgos would have had to contend with moth and sawfly caterpillars that gobbled down the florets and leaves during the night, leaving denuded branches the next morning. Cretaceous aphids, scales, planthoppers, and those long-beaked cicada-like insects, the palaeontinids, might have sucked the juices from these plants, causing the leaves to wilt or drop prematurely.[35] Small beetles that relished the pollen-rich male flowers had the ability to reduce the reproductive potential of these trees.

Cycadophytes and cycads were diverse groups in the Cretaceous and ranged from dwarf to colossal plants. Iguanodons and stegosaurs that indiscriminately grazed on the fronds probably disturbed scarab beetles devouring the leaf edges. The pollen and seed cones of these plants would be relished, although those that produced nuts with toxic chemicals may have been avoided—but not necessarily, since baboons in Africa have no problem dealing with the undesirable compounds in the nutritious cycad seeds.[57] Dinosaurs undoubtedly competed with guilds of insects feeding in the male and female cones, including small weevil larvae in the seeds. The starchy stem of cycads unquestionably provided food for pachycephalosaurs and hypsilophons, but they had to contend with longhorn beetle and moth larvae.[58] Other insect competitors on these plants possibly included sap-sucking aphids or scale insects that left layers of sticky, whitish residue on the dying leaves. And by the mid-Cretaceous, ants were beginning to distribute these sapsuckers throughout the forest.

Another abundant group of dinosaur food plants were ferns. These varied in size from dwarf maidenhairs and medium-sized bracken ferns to towering tree and seed ferns. In fact, some of the tree ferns probably grew in such dense clusters that they concealed the dining dinosaurs. The stegosaurs were no doubt quite fond of the tender fronds of these plants, but probably competed with sawfly larvae, moth caterpillars, weevils, and gall midge larvae living in chambers along the leaf edges. Aphids, plant bugs, leafhoppers, and planthoppers possibly vied with dinosaurs for this resource. A dinosaur delicacy certainly was their soft stalks and starchy, pith-filled trunks, both developmental sites for weevils. Some of these primitive weevils still exist, and one Australian species caused considerable damage to tree ferns when accidentally introduced into Hawaii.[59] Many of the Cretaceous ferns likely had edible rootstocks[60] that represented a delicacy for both protoceratopsians and beetle larvae.

Dinosaurs feeding on conifers encountered a host of insect competitors. Pachycephalosaurs can be envisioned eagerly gathering around araucarians when the cones dropped. Both the pollen and seed cones are highly nutritious, with elevated levels of carbohydrates, proteins, fats, and trace elements. Here the dinosaurs vied with weevil larvae developing in the pollen cones[34] and moth (Agathaphigidae) and beetle (Belidae) larvae in the seed cones[61] (color plate 12C). Today, caterpillars of *Agathiphaga* can destroy up to 95% of kauri seeds in a cone.[182] There may even have been gall wasps devouring the seeds[62] (color plate 5B). Diplodocids and titanosaurs feeding on leaves of young araucarian trees no doubt shared their meals with both leaf-mining and foliar caterpillars, as well as thrips, aphids, and scale insects (color plates 3A, 3E, 3F). The ancestors of a small moth (Gracillariidae) whose immatures are found mining the leaves of New Zealand kauri trees,[63] as well as those of giant weevil larvae that develop in araucarian trunks, may have been present at that time.[64]

There would have been fringe-winged thrips on the buds, flowers, leaves, and stems of all types of plants (color plate 3A). These small insects have mouthparts adapted for sucking and exhibit two basic feeding patterns, a shallow type mostly restricted to epidermal cells and a penetrating type that probes the deeper plant tissues. The feeding punctures, as well as the oviposition slits, cause mechanical injuries to plant tissue, and if the insects are numerous, defoliation can occur. This type of damage is caused by thrips that attack kauri trees in northeastern Australia.

Other important conifer defoliators in the amber forests were sawflies, many of which are gregarious. Today, the widespread red-headed sawfly is quite destructive to pines, larches, cedars, and spruces.[65] If the attacked trees are not killed outright, the damage can retard growth and kill off limbs and twigs. Other conifer sawflies that existed back then were leaf-feeding tenthredinids and web-spinning sawflies, as well as xyelids that bred in cones, buds, and developing shoots. Certainly, these insects were serious competitors with conifer-feeding dinosaurs.

All conifers as well as angiosperms were susceptible to attack by generalist feeders such as crickets, katydids, elcanids, monkey grasshoppers, shorthorn grasshoppers, wetas, and mountain crickets (haglids) (color plates 4A, 4B). From their fossil record, haglids were quite diverse in the Mesozoic, and their probable behavior can be determined in part from the species still existing in North America. One that feeds on conifer pollen has singing males that offer not just a song for the chance to mate, but their fleshy wing flaps as a tasty treat. The collective feeding habits of the Cretaceous grasshopper-like locustopseids and elcanids probably destroyed many plants, just as hordes of locusts strip crops now.[65] And short-horned grasshoppers must have attacked a variety of plants back then just as they do today. Insect defoliation of the conifers may have killed off many trees since they cannot produce new leaves and recover like angiosperms.[66]

Saplings of *Metasequoia* were certainly another preferred dinosaur plant. As sauropods devoured the tender leaflets, they possibly encountered leaf beetle larvae concealed between the needles, mealybugs within their white paper-like cocoons, and small caterpillars hidden in the bud tips. And the dinosaurs that selected low branches of cypresses shared their dinners with small bud caterpillars, sawfly larvae, and scale insects with their attending ants. Sap-sucking mesozoicaphidids, which appear to be related to modern adelgid aphids, were likely responsible for the decline and death of many conifers in the amber forests. When the hemlock woolly adelgid was introduced into eastern North America, it spread through one-third of the forests, killing many hemlocks along the way.[67] These xylem-feeding bugs inject toxins into plant tissues, causing the needles to turn yellow and fall. This makes the plants more susceptible to drought and attack by other insects and plant diseases. Because of the combination of wind dispersal (they are quite light) coupled with rapid population buildup by parthenogenetic forms, these insects were responsible for the widespread death of hemlocks in North America. Hemlocks die about four years after the initial attack, and when the mature trees are gone, the canopy gaps are taken over by the more rapidly growing angiosperms. A similar scenario is reasonable in the Cretaceous amber forests.

The enigmatic *Brachyphyllum*, an extinct Mesozoic gymnosperm, was quite widespread in the Cretaceous. Dinosaurs dining on the leaves and fallen pollen cones of plants would have encountered bizarre, elongate grasshopper-like insects as well as eccentric walking sticks and erratic sawflies, all of which have been recovered with the remains of *Brachyphyllum* leaves and pollen in their alimentary tracts.[35]

In 1962 an entomologist proposed that dinosaurs could have been eliminated by caterpillars that destroyed their food sources.[68] As their plants were consumed by hordes of moth larvae, dinosaurs died from starvation. While this theory emphasizes the competition that occurred between foliage-consuming insects and herbivorous dinosaurs, it is impossible to envision it resulting in any type of dinosaur extinction event.

Did some dinosaurs have a taste for mushrooms and other fungi? Did some hunt truffles? That is difficult to say, but if fungivorous dinosaurs existed, they certainly had their choice of food items. These included various types of mushrooms, bracket fungi, puffballs, and truffles (color plate 12A, fig. 12). A range of animals, from rodents and pigs to deer and banana slugs and even tortoises, lizards, and birds seek out fungi. But no matter what type was selected, whether gilled mushrooms, puffballs, or club mushrooms, they certainly came with an assortment of insect competitors, including a variety of beetles and flies.

# 5.

## Did Dinosaurs or Insects "Invent" Flowering Plants?

WHILE RELATIVELY few angiosperms were established at the beginning of the Cretaceous, by the Late Cretaceous flowering plants accounted for possibly half of the plant diversity. With their amazingly rapid growth rates and relatively short reproduction periods, these plants were predestined for success. Encoded in their genetic makeup was the ability to radiate into a variety of habitats, from bogs and marshes to stone crannies, mountaintops, and tree branches.

In his book on dinosaurs,[52] Robert Bakker felt that plant-eating dinosaurs could have "invented" flowering plants. He concluded that in contrast to the Late Jurassic browsers that fed on foliage in the canopy and subcanopy layers, Cretaceous dinosaurs were predominately grazers that indiscriminately clipped the flora to near-ground levels. Angiosperms, which grow and reproduce quickly, recovered from this clear-cutting faster than gymnosperms, thus giving them a competitive advantage that eventually led to their dominance. However, this theory has met with criticism[69] and Paul Barrett and Katherine Willis rejected Bakker's theory on the grounds that angiosperms were "neither sufficiently abundant nor widespread to have been a major component of dinosaur diets during the Early Cretaceous." It is also unlikely that most Cretaceous dinosaurs grazed the vegetation to the ground, and seedlings of both angiosperms and gymnosperms would have survived. Also, Bakker compared widespread dinosaur herbivory to that of mammals grazing on grasslands, and there is no evidence that such habitats occurred at any time in the Cretaceous.

Early Cretaceous herbivorous insects definitely played a significant role in angiosperm evolution through their feeding on archaic gymnosperms. Occurring in all strata, insect damage would have been found from the roots to the tips of the leaves. By stripping the spreading cycad, conifer, and fern foliage, devouring the stands of horsetails and carpets of club moss, and avoiding the angiosperms, they weeded out the competition, and their selective feeding worked in favor of the flowering plants. So if herbivory in the Cretaceous facilitated the diversification of flowering plants, the insects could certainly take most of the credit.

When insects with appetites for angiosperms appeared on the scene around the mid-Cretaceous, they defoliated the limbs, bored into stems, and fed on flowers and fruits. Subterranean creatures such as termites, beetle grubs, fly maggots, caterpillars, and aphids ate away on the roots and underground stems.[65] By selective feeding on specific genera and species of angiosperms, Cretaceous insects were determining which lineages of flowering plants would be here today.

Another activity, however, put insects far ahead of dinosaurs as the driving force behind angiosperm evolution, and it had nothing to do with herbivory but everything to do with plant reproduction. Insects became the prime transporters of genetic material from one flower to another, replacing the chance encounters of wind pollination with the personal transportation of gametes.

# 6.

## Pollination

*In a dark recess at the base of a fern, a tiny bee began the arduous duty of motherhood by constructing a nest. She first cleared a small area on the ground by using her toothed mandibles to remove, particle by particle, larger bits of soil, then began digging in earnest with both front and middle pairs of legs. As the small pile of excavated soil became larger, the busy animal gradually disappeared inside the tunnel until finally only the ejected soil particles revealed the hymenopteran's presence. Working day and night, the minute insect formed a cell that would serve as a nursery for its first progeny. She added the finishing touches by compacting the soil in the chamber with the tip of a blunt abdomen.*

*With the first task completed, this mother now had to concentrate on the perilous duty of filling the cell with pollen for any offspring. In the dim light of the following morning, the solitary bee prepared to depart. Before that first foray away from the nest, she flew up and circled the area three or four times to memorize the location and then whisked off through the forest. Winding a tortuous way between the huge araucarian trees with their glistening resin-spotted trunks and dodging the hanging lianas, the diminutive insect noticed some small, intricate white flowers wedged between clusters of blue-green leaves.*

*An assortment of climbing and hovering insects was already busy feeding on the blooms. Metallic-colored beetles with comb-like antennae devoured the pollen while long-legged flies mopped up nectar and miniscule thrips with flashing silvery wings scraped tissue from the petals. Avoiding these, the little bee settled on a group of recently opened flowers and began transferring as much pollen as possible onto the stiff hairs of her hind legs. Even with a full load, she would still need to make another five or six trips before collecting enough pollen to nourish just one larva to adulthood. Arriving back at the nest, the female quickly entered the nursery cell and began scraping off the pollen. As she left, she blocked the nest entrance to foil any thieving insects that might be searching for a free meal by partially filling the entrance hole with a plug of dirt and then placing some debris on top as camaflougue. Then it was back to the forest to search for more flowers.*

*The journey this time led to a clearing where there was already a tumult of activity, but not just from insects. Ceratopsians were grazing in the undergrowth, consuming everything in their path like lumbermen implementing a clear-cut operation. Browsing indiscriminately on ferns, cycads, horsetails, low-lying conifers, and shrubby angiosperms, the giant herbivores devoured the herbage and shrubs that had regrown since their last visit.*

*Oblivious to any competition, the hymenopteran went from flower to flower collecting pollen, stopping now and then for a sip of nectar. Her hairy body was covered with pollen and with each visit to a neighboring blossom, some of these grains were left on the female flower parts. The simple act of cross-fertilization would insure that seeds would be set and a new generation would appear. This was a gift delivered in exchange for pollen and nectar that the bee needed for procreation.*

*With hind legs loaded to capacity, the hardworking bee returned to the nest. Having now gathered enough provisions for one offspring, some nectar was added to the pollen grains and the mixture was shaped into a little ball. Care was taken to keep this resource from touching the moisture-laden earthen walls since an attack of mold could make it inedible for the young. When satisfied, the female turned and laid a pearly white egg on the top of the food mass, then hastily backed out of the chamber and closed the entry securely by tamping down a layer of soil particles over the hole.*

The most significant way insects aided the establishment and spread of flowering plants was by their pollination activities, something the dinosaurs were incapable of doing. Most of the early plants were wind and water pollinated, and neither dinosaurs nor insects took an active part in this process. However, this did not keep any number of insects, from beetles and thrips to flies and wasps, from feeding on pollen from cycads, cycadeoids, and conifers. Those ancient associations may have been strictly one-sided, with the insects just eating the pollen and leaving, but at the same time, some could have fortuitously transferred pollen from one plant to the other. We know that grasshoppers and sawflies consumed pollen from Mesozoic gymnosperms since they have been preserved with these grains in their guts.[70] Insects that habitually fed on the pollen of certain plants may have become the major or even sole pollinators of those species when the associations became more stable.

By the mid-Cretaceous, many pollen-eating insects added the angiosperms to their list of food plants. While the rate of insect pollination of mid-Cretaceous flowering plants may not have been as high as 70%, which is roughly what it is today, certainly a number of beetles, flies, wasps, and thrips would have dined on this resource.[71] Although many of these flower-visiting insects were chance pollinators, some plants undoubtedly depended on them. Probably the ancestors of belid, brentid, and molytine weevils and langurid beetles that currently fertilize cycads had already assumed that role in the Cretaceous.[48,72] The attraction offered these beetles are shelter, food, a place to mate, and a site for larval development. The belid weevils dine on the starch-rich structures (sporophylls) that support the pollen sacs, while the langurid beetles ingest the pollen directly. By thus partitioning the food resources, they can coexist on the same plant. Other types of cycads are pollinated by small molytine weevils whose adults feed on the pollen and larvae consume the starchy stems.

Other gymnosperms in the amber forests were probably pollinated, in part, by nemonychid, belid, carid, and brentid weevils, which today are associated with araucarians, cypresses, and podocarps. Even leaf beetles thought to have developed in the trunks of cycadophytes could have been pollinators.[35] It is likely that the Lebanese amber nemonychid weevil may have pollinated the resin-producing tree since they are presently known to develop in male cones of araucarians and other conifers.[34]

Current associations reported between weevils and palms possibly began in the Cretaceous. Masses of African derelomine weevils gather to feed, mate, and deposit their eggs on male oil palm flowers. There they pick up and later carry the sticky pollen grains to female flowers. This coevolutionary behavior pattern has become so established that mites and nematodes feeding in the palm flowers have come to depend on the beetles to transport them from plant to plant.[73] If the weevils suddenly disappeared, so would the mites, nematodes, and possibly the palms.

As angiosperms diversified, so did the pollinating insects. Early angiosperms competed not only with gymnosperms, but also with each other to win the attention of potential pollinators. They developed more tasty rewards, as well as visual and sensual clues in the form of attractive colors and fragrances to draw the insects. And as primitive moths, flies, wasps, and bees became more efficient, a greater percentage of flowers were fertilized and more seed was set and distributed, thus insuring that when new habitats became available, rapidly growing angiosperms were able to colonize them before gymnosperms.

The most dependable pollinators are those that use this resource to supply protein to nourish their young. They constantly visit a succession of flowers, depositing pollen on the stigmas. Having a messenger deliver the grains, even if some are sacrificed as food, certainly is a superior method than releasing masses of pollen and counting on the wind to distribute them to the right location.

The first steadfast pollinators were probably wasps that substituted plant protein in place of arthropod prey for their young. These wasps could have transported pollen on various parts of their bodies, possibly on their facial hairs, which is how an extant wasp carries pollen to her young.[74] The actual method of transfer presumably depended on the size and properties of the pollen (sticky or dry, clumped or separate, etc.).

At some point in the Cretaceous, hymenopterans adapted morphological features that facilitated the collecting of pollen, the most significant of which were branched hairs. Such plumose hairs served as pollen nets to retain the grains and made transportation to larval habitats much easier. Today, plumose hairs occur on all bees, and the transition from wasp to bee regarding this character apparently occurred in the Early Cretaceous. Just when bees appeared has been estimated by molecular studies to be about 125 million years ago.[75] These early "protobees," like the one in Burmese amber,[44] still retained a few wasp features (color plate 14B).

Cretaceous wasps and primitive bees spent a significant amount of time visiting and pollinating flowers, and the plants responded by making their flowers more attractive. Thus, showy and colored petals, fragrances, oils, and nectaries appeared, the latter not just positioned on or at the base of the petals, but along flower stalks, stems, and leaves. Such props would raise a flag and beckon, just like carnival barkers, "Come and visit me first."

At present bees, with thousands of species, are the most important pollinators in the world, and all depend on angiosperms for survival. Most bees are solitary, as was the Cretaceous *Melittosphex*, and collect pollen from specific plants or plant groups or from plants in special habitats.[76,77] They often stake out a home range within a specific ecozone like moors, dunes, meadows, or heaths. All tasks depend on the single female, which collects the pollen, carries it to the nest in the ground or decaying wood, stores the grains in cells, deposits an egg in each cell, and then seals up the chambers. Each egg develops into a larva whose food is limited to that stored with it, which usually amounts to just enough nourishment to reach adulthood.

Just when social bees appeared in the past is difficult to say because while the previously reported Cretaceous social bee (*Cretotrigona*) has been discredited,[78] primitive stingless honeybees could have evolved by the Late Cretaceous. Their well-developed societies and requirements for nectar and pollen would have made them efficient pollinators of flowering plants, just as they are in the warmer parts of the world today.[79]

The new insect pollinators emerging in the beginning of the mid-Cretaceous definitely gave the angiosperms a tremendous advantage. This meant that the flowers did not need to produce copious amounts of pollen to be distributed by wind currents. A small amount of pollen delivered by a dependable insect that had just visited the same type of flower was a much more efficient system, and *Melittosphex* is proof that this happened some 100 million years ago.[44]

Just how would these pollinators have affected the dinosaurs? If angiosperms were uncommon at the beginning of the mid-Cretaceous and restricted in their distribution, as has been suggested,[69] the dinosaurs would have fed mainly on gymnosperms. By increasing angiosperm diversity through pollination, insects would have given the flowering plants a competitive edge over the gymnosperms and dinosaurs would have suffered. However, in the Late Cretaceous, when at least some dinosaurs fed on angiosperms, pollinating insects would have assisted them by ensuring additional variety and the spread of flowering plants. Insect pollinators were keystone species that not only benefited the angiosperms, but all insect and vertebrate herbivores that depended on these plants for food, as well as predators and parasites that relied on the herbivores for survival.

# 7.

## Blights and Diseases of Cretaceous Plants

*Pterosaurs gliding above an expanse of the araucarian forest passed over a bronze swath that cut across the otherwise verdant landscape. Conifers within that reddish-brown patch stood denuded of foliage and their gracious branches had turned hard and brittle. All around the periphery of this stark graveyard of dead trees were diseased ones destined for the same fate, their foliage already turning various shades of yellow and red.*

*A little bark beetle scrambled over the trunk of a still-healthy kauri tree, looking for a place to bore through the bark and construct galleries for her brood. She passed similar beetles that had searched with the same zeal, but were now being suffocated by droplets of resin the tree poured out through their drilling holes. It was the tree's main line of defense against such attacks. But along with many others, this beetle was finally successful, and every completed mission was marked by scatterings of wood dust under a small entrance hole in the bark.*

*As soon as the female reached the tender plant cells under the bark, she began to eat away at the wood, instinctively constructing long, narrow tunnels punctuated with lateral egg galleries. The wood borings and frass left over from construction were shoved backwards into passages or dumped to the outside through small borings in the bark. At night the air was filled with a cracking sound emanating from the concerted chewing of bark and wood by thousands of such beetles. After finishing a main gallery, the beetle fashioned some lateral egg galleries where a few small white eggs* were deposited. While excavating, minute spores of a symbiotic fungus were released from her body. Within a few days, the fungus grew hyphae along the surface of all the tunnels and provided food for the young. Devouring the fungal-riddled wood, the legless larvae grew rapidly, and when their development was nearly finished, they built an enlarged pupal chamber just under the bark. Life appeared safe and secure in those galleries sealed off from the world of outside predators. But other beetles the same shape and size of bark beetles had entered some of the tunnels and ate many of the developing stages there.

Those pupae that escaped the marauders changed into soft yellow-brown adults and waited until their exoskeletons hardened before they chewed out of their domiciles. As they left, they carried in cracks and crevices on their bodies the spores of the symbiotic fungus that was so necessary for their development. The fungus infection they left behind in their abandoned homes produced a blue stain in the wood as the disease progressed and gradually spread throughout the tissues of the tree, sealing its fate. Leaves of the infested araucarian first turned yellow, then slowly withered and died until only a lifeless trunk remained.

In another part of the forest, a female siricid wood wasp was preparing to implant eggs into the stem of a conifer. She used a saw-like ovipositor to bore through the bark and then deposited about a dozen creamy white eggs in the deep recess, where they would be protected from predators. When laying eggs, spores and fungal fragments stored in compartments at the base of her abdomen were also released. Survival of the larval wood wasps depended on the growth of that symbiont which ironically would eventually destroy the tree. Over the course of the next few months as the succeeding generations of wood wasps spread the disease, large tracts of conifers yellowed, turned brown, and died.

Meanwhile along the riverbeds, palms dropped their fruits prematurely. Flowers turned black and fell off the stalks, while their leaves, beginning with lower ones, progressively turned yellow, then brown. Small, colorful leafhoppers feeding on the diseased *plants picked up the microscopic agents that were responsible for this viral disease and at subsequent feedings unknowingly distributed them far and wide. These plant diseases were an integrated part of the forest's life cycle.*

The Cretaceous was a moldy world, not that much different from the tropical regions today. Fungi parasitized other fungi, and these in turn were parasitized by still others.[345] Having lived in the tropics, we know what it is like to find masses of long gray filaments emerging from shoes left in the closet a few days, spots spreading over various parts of your skin, spores clogging your respiratory system, and delicate strands etching the surfaces of microscope lens. In fact, practically all of the microscopes in our West African laboratory were useless because of fungal-caused scratches on the lens surfaces that blurred the images. And it was particularly frustrating because it was not possible to simply clean the lens; you had to regrind them before they became operable again—not an option in an African field compound!

The range of habitats that fungi can fill is truly amazing. A random microscopic examination of the leaves and stems of plants gleaned from Cretaceous forests would have revealed a wide variety of fungi. Brown leaf-spot fungi peppered the tops of araucarian leaves while clusters of rust spores colored the bottom. Twigs of these dominant trees were covered with strands of white fungi, and other types of molds entered the stems, causing large, gnarled cankers through which other types of wood-rotting fungi entered. Even young seedlings suffered from damping-off fungi that destroyed their roots before they had a chance to reach a foot in height. If we were able to travel to a locality in a Cretaceous forest before and after fungal diseases had run their course, we would immediately notice a difference in the appearance of the plant community, brought about by the demise of very susceptible species and the resurgence of more resistant types.

Insects certainly had a great impact on the floral composition by carrying pathogens that attacked the dinosaurs' food plants. Based on our present knowledge, insect-vectored fungal and viral diseases were critical in determining which plants lived and died in the Cretaceous world. While bacteria and protozoa infect a few plants, these organisms did not develop the extensive pathogenic associations with plants seen with fungi and viruses.

Bark beetles are small, dark insects about the size of pinheads, but armed with their symbiotic fungi, they become harbingers of death and destruction. These insects bore through the outer bark of evergreen and deciduous trees to make brood galleries in the sapwood.[65,80] Unfortunately the same fungi that feed the beetles also kill the host tree by plugging its water-transporting cells.[81] Araucarians, podocarps, cypresses, and *Metasequoia* were certainly just as susceptible to bark-beetle attacks in the Cretaceous as they are now,[82] and some consider that Cretaceous araucarians were the first trees that bark beetles attacked.[83]

The devastating effects of fungal diseases carried by these beetles were made evident over the past hundred years when two pathogens arrived in North America. One was the notorious Dutch elm disease, which appeared in 1930 and within fifty years had spread from coast to coast. The efficient dissemination of the deadly fungus by two bark beetles essentially destroyed the American elm throughout North America.[84]

The second example was the chestnut blight. Introduced into the United States from Asia around 1900, it destroyed over 99% of the several billion American chestnut trees in the eastern forests within fifty years. Asian chestnut trees were susceptible to the infection but were not killed, while American chestnuts, never experiencing the malady, had no resistance. This is considered to be one of, if not the most, destructive plant diseases known. Wood-boring insects, including longhorn beetles, tumbling flower beetles (color plate 5D), and bark beetles contributed to the dispersal of this pathogen.[85]

Just when bark beetles acquired symbiotic associations with fungi is an interesting question. Bark beetle (scolytid) fossils occur throughout the Cretaceous and probably acquired the dependency on fungi early in their existence. Many extant bark and ambrosia beetles carry the spores and hyphal fragments of these fungi in special pouches (mycangia) on various locations on their body.[86]

Another deadly duo that feasibly occurred in the Cretaceous involves a wood wasp and its symbiotic fungus. While most wood wasps breed in dead timbers, *Sirex noctilio* selects living pines as breeding sites. In Europe where this insect is native, the pines have adjusted and usually do not die from the invasions. However the Californian Monterey pine, which New Zealanders and Australians grow commercially as a timber tree, is highly susceptible. The small, native coastal populations of Monterey pine had lived in splendid isolation from siricid wasps, but when the duo met in Australian plantations, it was chaos. Within ten years, 4.8 million Monterey pines were killed, and without a successful biological control program, the wood wasp and fungus would have eliminated the Monterey pine from that continent.[87] It is quite likely that Cretaceous wood wasps carried symbiotic fungi for nursing their broods.

Longhorn beetles carrying pine-wood nematodes and fungi show how complex and lethal insect-plant pathogen combinations can be. Some species of longhorn beetles inoculate both nematodes and fungi into pine stems when they lay eggs. The beetle young, as well as the nematodes, consume the fungal-infested wood, but the nematodes spread and clog the water transporting system of the trees. Infected trees die within 30–40 days after showing the first symptoms. This disease was very serious in Japan in the 1940s, causing annual losses of about 400,000 cubic meters of timber over the following 25 years.[88] The native Japanese pines are highly susceptible, while North American pines, where the disease may have originated, have some resistance. Longhorn beetles were present in the Cretaceous and nematodes resembling the pine-wood types occur in Lebanese and Burmese amber.

Another deadly group of fungal parasites are the rusts, with both beetles and flies serving as vectors. White pine blister rust is one of the most important forest diseases in North America, and it would be nearly impossible to grow pines without controlling the pathogen.[89] While our white pines were saved by fungicides and selecting resistant trees, would naturally occurring plant chemicals have been enough to save Cretaceous conifers from similar pathogens? We can only guess how serious new strains of rust fungi could have been to archaic gymnosperms. A present-day rust fungus requires both ferns and firs to complete the life cycle, and similar associations between ferns and conifers certainly existed in the Cretaceous.[90]

Kauri trees are susceptible to attack by a number of fungi,[182] and we assume that similar types invaded Cretaceous araucarias. We know that various types of mushrooms in Burmese amber were growing on the bark of araucarians[91,345] (color plate 12A, 12D). Some of these kinds, such as the cauliflower mushroom, cause root and trunk infections on present-day conifers[92] while others are pathogenic on angiosperms.[93] Insects found in the amber together with an ancient club fungus possibly distributed the spores just as beetles carry mushroom spores inside their guts today.[94] If pathogenic, the Burmese amber club fungus certainly affected at least one of the dinosaur's major food plants.

A combination of extremely susceptible trees, relentless insect vectors, and novel pathogens provides the impetus that extirpates plant populations, and if fungi were establishing new symbiotic associations with Cretaceous insects, scenarios like those described above may have been widespread.

We turn our attention now to insects vectoring plant viruses and mycoplasms (small single-celled organisms) in the Cretaceous. Did dawn redwood leaves turn yellow, fern leaves show mosaic patterns, or araucaria leaves have ring spots? Was the landscape dotted with stunted cycads or tree ferns displaying wrinkled fronds? Had some of the early angiosperms suddenly turned yellow and wilted? If so, the cause could have been virus infections.[89]

Aphids, leafhoppers, planthoppers, whiteflies, and in fact all of the sucking and even some biting insects that lived in the amber forests had the potential to carry plant viruses (color plates 2 and 3). It has been proposed that associations between viruses, insects, and plants coevolved some 200 million years ago, long before our earliest amber site in Lebanon.[95]

Cretaceous aphids were quite diverse, and considering that over three hundred viruses are borne by them today,[95] we assume that they were transmitting these pathogens back then (color plates 2C, 3E). Some of the earliest aphid-transported viruses could have infected ferns.[96] When virus-infected aphids finish feeding and withdraw their long, slender mouthparts from the plant tissues, they close the feeding site with a salivary plug tainted with these pathogens.[95] These viruses can kill their plant hosts, and while those infecting conifers and other gymnosperms have been only cursorily studied, they probably occurred in these plants groups in the Cretaceous. And as aphids developed vector associations with these infectious agents, their transfer over to angiosperms was only a matter of time. Palms are a group of early angiosperms that were probably infected with viruses, possibly belonging to the same group (potyviruses) that are vectored by aphids to African oil palms.[97] From our knowledge today, it is likely that insect-borne viruses played a large role in shaping the angiosperm's world.

Leafhoppers and planthoppers (color plates 2A, 2B, 3D, 3F) are other insect groups in Cretaceous amber that are known to spread plant viruses and mycoplasms. In fact, there are over 130 leafhopper species in 8 subfamilies that currently transmit these pathogens, which multiply in both plants and insects.[95] Many extant planthopper-borne viruses infect grasses,[95] so some of the representatives in Burmese amber possibly were disseminating viruses to early bambusoids as well as other flowering plants growing then.[17,18,348]

Mealybugs, present throughout the Cretaceous,[98,99] are now widespread in the tropics and subtropics (color plate 3F). These small sucking insects can kill plants just with their toxic secretions, but many carry viruses.[100] While some have suggested that

Mesozoic scale insects lived only on gymnosperms[98] and therefore could have introduced viruses into cycads, caytonias, and conifers, certainly many began feeding on angiosperms then and transmitted viruses to these plants as well. A condition known as swollen shoot disease of cocoa appeared when cacao was introduced into Africa from its native South America. A mealybug transferred a virus from native African plants to the newly introduced cacao plants, and many trees died or suffered extensive damage from this novel disease. This is yet another example of how a plant can be eliminated by an insect-vectored virus.

Whiteflies, members of the family Aleyrodidae, occur throughout the tropics and subtropics and were well represented in the amber forests. They are quite small, ranging up to only about 3 mm in length, and can be recognized by the white, powdery deposit on their wings (color plate 3B). The flat, oval immatures are usually found on the lower leaf surfaces where their long stylets suck juices from the host. When populations are high, feeding not only causes leaf damage and stunting, but the sticky honeydew that is excreted on the leaves clogs the plant's breathing pores and provides nourishment for sooty molds that interfere with photosynthesis. Whiteflies carry viruses to angiosperms and gymnosperms,[101] causing yellow mosaic patterns on leaf surfaces and veins and various types of foliage distortions. Cretaceous whiteflies likely transmitted viruses that infected ferns, ginkgos, conifers, and angiosperms.

Six economically important tropical plant viral diseases are known to be vectored by insects.[102] These cause swollen shoot in cacao, mosaic patterns on African cassava, streaks on South Africa maize, yellow mottling on Kenyan and Philippine rice, and bunchy tops in Southeast Asian bananas. They are all vectored by either leafhoppers, chrysomelid beetles, whiteflies, mealybugs, or aphids. All of these insect groups were present in the Cretaceous, and some certainly disseminated viruses to both gymnosperms and flowering plants.

It is interesting that the most destructive plant diseases appear when exotic pathogens enter established ecosystems or when introduced plants acquire infections from diseased endemic species. Native insects play crucial roles in spreading diseases in both cases. Disease outbreaks certainly occurred in the amber forests when sap-sucking insects were just beginning to establish associations with plant viruses. The susceptibility of Cretaceous plants to viral agents would have influenced the composition of plant communities.

So by 100 mya, plant pathogens vectored by insects had certainly begun to decimate some of the archaic gymnosperms that dinosaurs depended on for food, thereby affording new opportunities for emerging angiosperms. Could a vector-borne plant pathogen have been responsible for the elimination of ancient araucarians from the Northern Hemisphere? Most araucaria species now grow in a few relictual areas in South America, northeastern Australia, New Zealand's North Island, Indonesia, and some remote islands such as New Caledonia and Fiji. Were these island populations spared from insect-vectored diseases that eradicated this tree family from the Northern Hemisphere? New Caledonia is an isolated, remote area, yet it contains 5 species of *Agathis* and 13 species of *Araucaria*, representing 45% of this family's worldwide diversity. Tragically, after surviving for eons in splendid isolation, these New Caledonian populations have been mostly eliminated by habitat destruction, and over 80% of the species are now seriously threatened.[103]

The disappearance of gymnosperm and angiosperm lineages towards the end of the Cretaceous would have caused serious problems for those dinosaurs that depended on them for nourishment. This is just another fundamental example of how small and seemingly insignificant insects affected dinosaurs indirectly by vectoring plant pathogens that eliminated their food source.

# 8.

## The Cretaceous: Age of Chimeras and Other Oddities

*Among a clump of conifers, one species of shrubby angiosperm had crowded out its neighbors, and a number of curious insects had collected on the flowers. Small solitary hymenopterans were actively skimming back and forth over the stamens, stopping momentarily to scrape pollen onto bristly hairs covering their hind legs. The branches on those hairs helped retain the grains, although many dislodged and fell on the receptive female flower parts as these primitive bees went about their everyday business. One tiny hirsute body was actually shaking from the exertion of collecting, and this bee stopped momentarily to use the back two pairs of legs to wipe excess pollen from the wing surfaces while employing the cleft claws on the front legs to brace herself. Then the exhausted female flew to the hidden nest tunneled deep within the soil where she deposited the last brood ball of the season.*

*The mature kauri trees in the forest were each a microcosm teeming with arthropod activity. A tiny brown ant-like stone beetle crawled slowly over the many fissures on a trunk and eventually lowered an extended abdomen, inserted a pointed ovipositor under a loose piece of bark, and deposited several eggs. The small female then resumed her other everyday activities, which included hunting down that favorite food, mites. Scurrying over the tree surface, the ant-like insect froze when a mite crossed nearby, but it was not the preferred kind and she continued to search. Finally an-*

other type of mite appeared, one adapted to the crannies and crevices on the araucarian tree. A quick pounce secured the prey between the vise-like grip of strong front legs, made all the more efficient by their extra segment, a unique character that distinguished this small insect species from all others in the forest. The minute beetle voraciously consumed the small arachnid and then resumed patrolling the bark surface.

Much lower on the same tree, a tick larva waited patiently on a twig for a passing vertebrate. This creature desperately needed a blood meal since stored reserves were nearly depleted. A slow-moving dinosaur approached and the tick suddenly became alert, extending the palps and front legs in anticipation. The outstretched palps were armed with a cluster of claws near their tips, a feature that made this particular tick different. As the dinosaur unknowingly brushed against the twig, the opportunity had come and the tick half-dropped, half-scrambled onto the moving animal, having completed the first important step in procuring a meal.

Up in the leafy crown of the same tree, a group of aphids prepared to abandon their home turf on the foliage and search for other feeding sites. The females were filled with small elliptical eggs that were the stock to establish new colonies. Most had withdrawn the long, narrow stylets through which they took sap from the host tree. These aphids differed from other aphids because they possessed only a single pair of wings instead of the normal four. Some began to test their wings by extending them as far as they could. A barely detectable gust of wind brushed across the outstretched appendages and they became airborne, flitting their way upward. Some landed against sticky pools of resin while others eventually settled on some leaves silhouetted against the morning light.

Further down at the base of the tree, the dim rays of light faintly illuminated a group of small flies viewing the world in a unique way—upside down! Equipped with spines on their hind legs that faced backwards, some females just dug a single prickly leg into bits of moss, while others attached both legs to rough bark fragments. Their clear wings contrasted with the dark bark as they hung mo-

*tionlessly. These vampire-like flies, which were equipped with uniquely-shaped serrated mandibles, were slowly digesting the blood they had obtained during the night.*

I was astonished when I peered down the microscope lens and saw my first Cretaceous chimera. When studying insects in younger Dominican, Mexican, and Baltic amber, it was commonplace to place the fossil in a modern family and sometimes in an extant genus. But these mid-Cretaceous fossils opened up an exciting, strange new world with creatures bearing combinations of characters never before seen among the living. Besides instilling wonderment, trying to identify these enigmas kept me awake at night since according to established standards, they shouldn't have existed at all. It made me realize that the current classification system has little significance at and below the family level for insect fossils 100 million years old.

These were reminiscent of the mythological she-monster with a lion's head, goat's body, and serpent's tail that the Greeks called a chimera, and that term is adopted here for fossils with morphological features found in two or more present-day groups. And for that very reason, knowing or deciding how to classify them is a conundrum. Vertebrate paleontologists have *Archaeopteryx*, a strange animal with teeth and feathered wings with claws that appears both bird and reptile. We entomologists have more numerous and just as interesting chimeras, proof that the Cretaceous was an age when species radiated extensively both structurally and behaviorally, and the partitioning of habitats was continually changing.

Because attempting a DNA analysis of these rare fossils necessitates destroying them and therefore would not be feasible, our conclusions are based on their physical features. We are left with speculating on how they arose and why they became extinct. Some of their characters could be the result of convergent evolution, where two entirely different lineages acquire the same features. Others are possibly due to spontaneous mutations followed by adaptive radiations, and a number might be throwbacks to earlier lineages where inhibitors were mistakenly switched off during development and latent DNA in the genome activated. A few may have the characters of two groups because they are intermediate between two lineages.

My first Cretaceous insect chimera was the little Burmese amber *Hapsomela*, an ant-like stone beetle with both front legs equipped with 6 instead of the normal 5 segments found in all living insects (fig.16).[39] How astonishing that only the front pair of legs had the extra segment while the remaining had the normal number! I kept staring at those front legs, looking at each segment over and over again, not wanting to accept the obvious. Should I just place the specimen back in the container and file it away? No, this was worth presenting to the scientific world even though the facts went against all conventional wisdom. The closest modern groups that have extra-segmented legs are spiders and mites, and this fossil certainly didn't fit there. Why would any insect need additional leg segments? While crustaceans, arachnids, and millipedes have more, modern insects do quite well with their basic five. However, we do know that some ancient insects found extra joints useful because a number of Paleozoic forms had an extra one or two.

I began to contemplate how this chimera arrived on the Cretaceous scene. Did the lineage originate in the Permian some 300 million years ago, continue undetected by paleontologists into the Cretaceous, and then disappear? That would mean the little beetle was one of the terminal descendants of an ancient group. Or was this a throwback, suggesting that insects retained the capacity for making six-segmented legs in their genome. If so, then something switched off some inhibitors and activated a long-dormant segment of DNA, making those segments appear. Maybe this individual was just the result of a single spontaneous mutation that happened to get captured in amber. But what would be the odds of a lone mutant becoming entrapped in resin and ending up under a scientist's microscope?

The tick *Cornupalpatum*, also from Burmese amber, likewise falls into our category of chimeras. A quick inspection suggested that it was just a typical hard tick (color plate 11E). However, a more detailed examination revealed that the pair of small sensory appendages known as palps had 5 terminal claws.[45] While claws are found in that location in some predatory mites, they have never been observed on present-day ticks. Here is a mite character on a tick, leaving us to wonder why the fossil came to have these talons and if their purpose was the same as that in predatory mites, subduing the prey. We are left to postulate whether they helped affix *Cornupalpatum* to its host, possibly some type of dinosaur, or were used to scarify mucous membranes in preparation for a blood meal.

So far, all Burmese amber aphids examined have only one pair of functional wings, while all other extant and extinct aphids have two pairs.[43] So while most characters orient these creatures with extant aphids, their vestigial hind wings align them with scale insects, whose males have rudimentary hind wings called hamulohalters (color plate 3E). Here again, we have chimeras with morphological features of two groups. What function, if any, could these stumpy hind wings serve? Flies (Diptera), which also have only two functional wings, vibrate their minute, residual hind wings (called halters) during flight, and some believe they serve as balancing organs. Perhaps these homologous structures on aphids had the same function or maybe they were just useless, vestigial organs on the way to oblivion, like the human appendix.

Since two specimens of the upside-down hanging flies, *Dacochile*, have been recovered, they probably were fairly common at the Burmese site. A debate centers around whether this little fly belongs to the moth fly family or primitive cranefly family.[46,47] The wing shape and venation are like those of primitive craneflies but the clear wing membrane and absence of a distinct wing lobe are moth fly characters. However, in toto, they actually have more features in common with the former, so they have been classified with the primitive craneflies (Tanyderidae) (color plates 10A, 10B). Problems arise when we attempt to determine what *Dacochile* was feeding on in the forest. It has well developed mandibles with serrated edges, indicating they were used for slicing or cutting into something, but what? Perhaps skin, maybe that of a vertebrate—in which case, dinosaurs would have been possible hosts.

We have already spoken about the small "protobee," *Melittosphex*, which contains characters found in both modern bees and hunting wasps (color plate 14B). With a body covered in plumose hairs, rows of hairs on the hind legs, and cleft claws, this hymenopteran had features of many bees. But with a pair of apical spurs on the middle legs, it also resembled wasps. Since this insect did not fit comfortably with extant bees or wasps, a new family had to be erected. Although small, this insect was probably quite efficient pollinating flowers, and was definitely a prototype bee.[44]

While chimeras have characters that resemble those in two or more modern families, there were some insect fossils that were just plain oddities. As far as can be determined from their remains, most of their features were unique. One example, a strange flea-like creature known as *Strashila* that roamed the earth some 150 mya,[104] could not be placed with certainty in a current insect order (fig. 21). We can imagine that this unusual specimen lived a furtive existence, resting under some debris or buried in the plumage of a bird or even a feathered dinosaur. With such huge, protruding eyes, detecting quarry was easy, and when suddenly extended, the enlarged hind legs would have effortlessly propelled the body high into the air and hopefully onto an unwary victim. After burrowing through the feather layer to the skin, the stout beak was inserted into the vertebrate and blood sucking began. The strange, paired appendages extending from the abdomen probably anchored *Strashila* firmly in place. This aberrant fossil has been categorized with the fleas, and while possessing some flea-like characters, most of its features do not resemble those of any modern insect, earning the creature an appropriate specific name, *incredibilis*.

FIGURE 21. The strange *Strashila incredibilis* lives up to its specific name by being one of the most enigmatic insects from the Mesozoic period.[104] Its large hind legs are adapted for jumping and would have grasped the shafts of feathered dinosaurs. Its proboscis could have been used for taking blood.

Questions about the nature of chimeras and oddities, such as their lifestyle, why and when they disappeared, and how should they be classified, will challenge paleobiologists far in the future. They certainly represent some of the most intellectually challenging facets of paleontology.

# 9.

## Sanitary Engineers of the Cretaceous

*On an alluvial flood plain near the amber forest, a large or-
nithopod dinosaur momentarily stopped grazing among the fern
fronds, raised its tail, and defecated. Before the last of the waste
even reached the ground, the air was crackling with the vibrating
wings of thousands of dung beetles of all sizes and shapes. From
large and round to small and oval, the beetles landed directly on or
adjacent to the evacuated material. Not a moment was lost, and
some began burrowing into the pile of waste even before the dinosaur had moved away.*

*In order to get a head start, a few small brown females began
laying their eggs immediately in the mound of partially digested
plant material. Their robust cream-colored larvae would dwell in
the dung heap for several days, hopefully developing fast enough
to avoid predacious rove beetles and slender parasitic wasps
that would eventually be searching for them, along with pecking
birds, scratching lizards, and other vertebrate predators.*

*A few dung beetle larvae ate some eggs of a parasitic nematode
worm that had lived in the dinosaur's intestine. The nematodes
would remain inside the beetles for life but could only finish their life
cycle if another dinosaur came along and gobbled down the infested insects.*

*Some of the larger dung beetles were more protective of their
progeny and after reaching the bottom of the pile, continued tun-
neling into the ground. No eggs were laid until the females packed
the tunnels with dung fragments dragged in from above. The developing larvae would feed on these brood balls in relative safety*

*since if the dung pile was broken apart by predators, the buried brood was hidden from view.*

*Other beetles never entered the dung mass, but broke off and molded small portions into balls that could be rolled far away from the original source. Using massive jaws to clip away a small portion of dung from the original pile and their front legs to shape the prize into a ball, the beetles turned around and used the two back pairs of legs to push the sphere along the ground. A latecomer suddenly appeared and attempted to steal a ball away from one of the females. She viciously defended her future brood cell, grabbing the opponent with her mandibles and flipping it over. After successfully fending off several additional marauders, the exhausted female dug a hole into which she rolled her prize, deposited an egg on the fecal mass, and buried them together by refilling the hole.*

*Soon mites and nematodes that had hitched rides on the adult beetles would begin multiplying in the remains and moving through and over the dung pile, encountering various stages of flies and beetles that were finishing their development and would provide transportation to yet another fecal deposit.*

*A group of immature garudimimids, looking like horny-crowned cassowaries, moved out of the forest and onto the plain under the watchful eyes of their parents. When they encountered the ornithopod dung, they paused and began to scratch through the mass, using their hind legs, sending feces flying while pulling the pile apart. The strong claws at the end of each toe sliced through the soft fibrous material and exposed scurrying adult beetles and unwary grubs which they snatched up in their toothless beaks. After only a few minutes of feeding on the exposed plain, they rushed off to seek cover in a nearby patch of shrubs.*

*Deep in the forest, an old, infirm dinosaur took a final breath and slowly sank to the ground, leaving his body to the whims of nature. In less than a minute, several large carrion flies that had detected death in the air landed on the still body and began depositing masses of eggs on the face, especially around the mouth opening.*

*Within the day, white legless maggots hatched from those eggs and began squirming into the mouth cavity, creating a hot, fetid environment that other insects couldn't tolerate. Eventually, the wriggling larvae finished feeding and left the cadaver, crawling some distance away before burrowing into the soil to pupate.*

*As the corpse dried, small brown beetles began feeding on the skin, dried flesh, and even on the surface of some of the bones. The eggs these oval insects laid hatched into minute larvae with long, stiff hairs that served both as camouflage and defense from predators on the lookout for insect prey. These beetle larvae consumed all the remaining flesh over the next few days, even the desiccated strips flattened against the exposed bones. After a short pupation period, the adults emerged, mated, and the females sought out new cadavers.*

The world in the Cretaceous would have been a fetid mess without insects. Can you imagine putrefying dinosaur corpses littering the landscape, heaps of dung remaining for months on end, and dead vegetation taking forever to be recycled? Eventually microbes, earthworms, and other scavengers would have decomposed this waste material, but certainly without the assistance of insects, parts of the prehistoric world would have been almost uninhabitable for many animals.

Dinosaurs, like a majority of animals today, voided solid waste. Certainly "waste" is a poor term since animal dung actually contains enough nutrients to nurture a wide range of smaller creatures like beetles, flies, earthworms, nematodes, and even miniscule mites. Defecation is a normal, everyday physiological act, although the end product may serve other functions. For example, hippos, rhinos, and quite a few other animals will scatter their dung, sometimes even scraping it into the ground, as a territorial marker, and probably many dinosaurs did the same. Once on the ground, feces from terrestrial vertebrates is processed by insects and a variety of other organisms. Waste disposal by arthropods is quite a complex process, especially when large amounts are deposited. For example, in the case of cow-pats, a succession of insects are attracted to the material, with each group preferring dung at a particular stage of "ripening",[105] just as there are people who prefer aged cheese over fresh.

Most coprophagous insects are beetles and flies, many of which are represented in Cretaceous amber. These insects often bring along mites, nematodes, and other creatures on their bodies, and sometimes parasites within. Further, some drawn to a heap are not interested in the product as such, but have come to prey on others feeding there. So dung is a microcosm containing quite a hodgepodge of creatures, each with their own agenda but all playing important roles in the breakdown of this resource.

To illustrate how important a balanced dung ecosystem is and why it must also have been functioning millions of years ago, one can look to Australia. Native dung beetles had become adapted to relatively small and specialized kangaroo and dingo feces. Everyone was surprised when a sanitary problem arose after introduced cattle populations reached commercially sustaining levels and began producing more solid waste than beetles could handle. The cowpats began to pile up, eventually paving over many of the pastures, and the situation became intolerable. This imbalance benefited the bush flies since they essentially had the feast to themselves, and they took full advantage of the situation. The result was huge fly populations that, curiously enough, became very attracted to people's faces.[106]

Needless to say, a considerable amount of money and manpower was invested to eliminate this problem by searching for dung beetles throughout the world—and not just any beetles, but only those that dined on cow waste. Finally, candidates were found in Africa and introduced into the southern continent. A portion of these adapted to Australia and settled in to do their business. Over time, the cowpats receded, the fly populations dropped, and the Australians were no longer bombarded by pesky flies. (In North America, some dung-breeding flies, like the face fly, are attracted to the heads of grazing animals. Certainly, dinosaurs also had their own types of face flies.)

It doesn't take much to imagine that similar beetles served as sanitary engineers for the large herds of herbivorous dinosaurs. And what a monumental task that must have been! Recognizable coprophagous beetle types had evolved at least by the Jurassic, since *Geotrupoides* occurred in Europe at that time and this genus was also recovered in Asia during the Cretaceous.[35,107] Dung beetles in general are too large to be commonly entrapped in amber, and most of their fossils occur in sedimentary rocks.

A number of Cretaceous scarab beetles could have fed on dinosaur dung.[108] The larvae of undetermined adult scarabs in Burmese amber (color plate 1A, 1B) may well have developed on dino feces. These adults were associated with fruiting bodies of a club mushroom also preserved in the amber,[91] which suggests that they had come to this site for feeding, a typical behavior of mature dung beetles.[106]

Certainly quite a few coprophagous beetle species that escaped fossilization or have not yet been discovered were around then feasting on dinosaur droppings. Those that have been found represent the three methods of breeding behavior found in extant dung beetles, namely the dwellers, the tunnelers, and the rollers.[106] The dweller types feed, oviposit, and develop in excrement, without bothering to make any nest at all. The tunnelers and rollers are the real engineers since they make dung chambers for their offspring. Tunnelers make shafts through the pile that extend to various depths into the ground below. These are then turned into nests when they are packed with feces fashioned into molded brood balls for larval development. The rollers are the showiest and most well known of the coprophagous beetles. They take a portion of the dung, shape the feces into a sphere several times larger than themselves, laboriously roll the ball away from the source, dig a hole, and bury it with an egg. At any rate, beetle fossils such as *Geotrupoides* and *Cretogeotrupes* were probably tunnelers, while *Proteroscarabeus* and *Cretaegialia* could have been dwellers and *Holocorobius* possibly a roller.

While few giant herbivorous reptiles are around today, there are many large mammalian herbivores, and ecologically speaking, elephants certainly are the largest and probably the closest we have for a comparison to dinosaurs. Elephants and some large herbivorous dinosaurs probably shared a similar type of fermentative digestive system, with both feeding on a variety of plants including trees and shrubs, and producing large droppings composed of only partially digested wet plant material. A mature elephant, voiding up to 22 pounds 17 times a day, can produce over 300 pounds of feces in 24 hours.[109] Up to 16,000 dung beetles have been reported to rush to just a small 3-pound dropping, scurrying this way and that in their haste to obtain their share.[110]

The ecology of dung removal was probably not that much different back in the Cretaceous but definitely operated on a considerably larger scale. Dinosaur herds had to be accompanied by hordes of beetles in the same way that elephants are today.[111,112] To extrapolate how much solid waste the largest known dinosaurs probably produced, compare sauropods to the African elephant. A mature male elephant weighs in at 6.5 tons, while a female may exceed 3 tons. Contrast that with estimated weights of 55 to 100 or more tons for the largest of the sauropods.[113,114] Taking the lower end of the scale still makes them 9 times larger than a male elephant, and as such they could have eliminated well over a ton and a half of dung per day. Then consider that sauropods, like elephants, traveled in groups some, if not all, of the time.

We know that elephant herd size varies seasonally with related units coming together in the wet season to form groups of 50 or more animals. It seems unlikely that such large assemblages would have been the habit of sauropods simply because of the limitations of available forage needed to sustain them. But even if the herd was only 10 individuals, that still means a minimum of 15 tons of feces scattered in their wake each and every day. The sheer numbers of just beetles processing such a prodigious output is almost unimaginable, possibly in the order of

200–250 million. Of course, many sauropods were only elephant size and as such probably formed larger herds. The elephants of the Tsavo game park produce 1,500 tons of dung per day, perhaps with an accompanying 24 billion beetles!

The largest dinosaur coprolite recorded was presumably that of the immense theropod *Tyrannosaurus rex* and measured over 17 inches (43 cm) long.[115] Analysis showed that these theropods ingested bone fragments along with the flesh of their victims, but there was no evidence of insect associates. However, another large dinosaur coprolite, a foot in circumference and almost the same in height, had dung-beetle burrows in it.[116] The fossilized scat was composed mainly of undigested conifers. This discovery shows that specialized coprophagy was well established during the reign of the dinosaurs.

All of those beetle species associated with dinosaurs are certainly now extinct. Today, most dung beetles prefer mammalian to reptilian feces, probably because there are very few herbivorous reptiles around, and they do not produce the amount of waste seen with herbivorous mammals. There are fewer coprophagous beetles associated with the droppings of carnivores than of herbivores. Herbivore excrement is probably preferred because of the high moisture content and abundant partially undigested plant material, which supplies copious nutrients for the beetle grubs.

There are, however, a few coprophagous beetles that have an appetite for reptile feces, possibly carried over from the distant past. Some seek out the droppings of plant feeders like land tortoises and iguanas.[117,118] In Turkey, one such insect lives only on the dung of the land tortoise,[119] and similar associations occur with the beetles *Copris* and *Onthophagus* in Alabama, Mississippi, Florida, and South Carolina.[120, 121] The adults search for the burrows and nests of gopher tortoises and lay their eggs in the soil beneath them. After hatching, the young feed on the tortoise dung.[122] It seems likely that such an interaction could have taken place under the nests of hadrosaurs and other dinosaurs. Since there is data suggesting that some dinosaur parents fed their young at the nest,[114] their accumulated waste would have been a source of food for beetles.

Scarab beetles are attracted to carnivorous reptilian dung. Two of these search out feces of the boa constrictor in tropical America,[117] and fresh lizard droppings attract others in Costa Rica.[123] So it's easy to imagine that beetles back in the Mesozoic were also breeding in the waste of carnivorous dinosaurs such as *T. rex*.

Dung beetles are not the only decomposers that would have been associated with dinosaur waste. We know that they presently compete with flies. While less conspicuous than the beetles, dung flies represent a widely diverse group capable of breeding in all types of animal feces.[105] Some cow-dung forms represented in Cretaceous amber are phorid flies (color plate 1D), moth flies, midges, and fungus gnats. Since the larvae characteristically breed in excrement with a high moisture content, they do not make tunnels that persist and can be later fossilized. Aside from finding cast skins of the larvae, which is difficult enough in recent droppings, it is almost impossible to find evidence of fly activity in coprolites. Although the highly chitinized pupal cases would be good candidates for fossilization, the fly larvae crawl away from the dung pile when they are ready to pupate. Thus, their pupal cases would not be preserved with the coprolite sample. This leaves amber as the main source for verification of coprophagous flies in the Cretaceous.

A parade of other invertebrates visits dung. Many are nonspecialized, such as springtails, earthworms, termites, millipedes, mites, and sow bugs, which prefer theirs dry and aged. To complicate matters, arthropod guilds come to parasitize or feed on the feeders. These include parasitic wasps, predatory dance flies, soldier flies, flower flies, rove beetles (color plate 1C), and others, all forming one active conglomerate in the dung world.

Furthermore, besides having competitors and parasites to worry about, there are vertebrate predators that relish dung insects. After all, their sluggish nature would make them easy prey, especially when they are competing with others at a fresh dropping. Today, various mammals and birds eat beetles in elephant droppings. Certainly in the Cretaceous, there were dinosaurs that dined upon similar insects.

Aside from disposing of dung, insects play an important role in ridding the world of dead animals. Just turn over a gull on the beach or a squirrel on the side of the road and you will see insects working their way through the dead animal, transforming decaying flesh into a mass of aggressive, squirming maggots. Various insects are sequentially attracted to carcasses at various stages of decomposition. In fact, the time that insects arrive at a corpse is correlated so precisely with the state of decay that forensic entomologists are often called upon to use insect stages to determine the time of death.[124] The flies are the first to arrive. Their young are particularly tolerant of a putrid, partially liquid diet found in the early stages of decomposition, while the beetles usually arrive much later after the carcass has dried out.

What would have been the sequence of events following the mortal attack of an injured ceratopsian by *Tyrannosaurus rex*? The giant herbivore, weighing 3 tons, would probably have provided meals for a variety of smaller dinosaurs after *T. rex* had finished with the carcass. But even as *T. rex* was feeding, the insects were on the scene and beginning their job. They would have swarmed over the raw flesh and blood, buzzing away as the tearing teeth of *T. rex* approached, only to resettle the moment his head withdrew. The flies probably alighted on the blood-drenched faces of the theropods and sought out fleshy fragments of the meal. Certainly, dinosaur bodies attracted large numbers of flies and beetles, many of which were the ancestors of those found today on dead vertebrates, including reptiles.

In a study of necrophagous insects attracted to a dead leatherback turtle in French Guyana,[125] vultures set the stage by first removing the eyes of the dead reptile. The resulting open wounds attracted flesh flies (Sarcophagidae) and blowflies (Calliphoridae). Additional damage by vultures resulted in more wounds that this time attracted anthomyid and muscoid flies. Cockroaches arrived at night and scavenged on the carcass (color plate 6D). These same insect types undoubtedly were drawn to dead and dying dinosaurs. We know that blowflies, which are one of the most common necrophagous insects today, occurred in the Mesozoic,[35,126] and as such, they must have deposited eggs around the eyes, nostrils, and mouths of dead or dying dinosaurs.

Moth flies (Psychodidae) are another group that breed in decomposing bodies, especially under warm, moist conditions. These flies commonly occur in Cretaceous amber and probably bred in dinosaur corpses along with other fly groups such as phorids and sphaerocerids.[124] Developing fly larvae would have attracted predatory insects such as ants that decreased the fly population by taking maggots back to their nests to feed their young.

After a cadaver has been stripped of the soft tissues by the first arrivals, a second group of insects enters the scene to demolish the remains. Especially obvious are the skin beetles or dermestids. Both the adults and larvae relish dried skin and tissues, and leave their signature in the bones that appeal to them as a substrate for their pupal chambers. Bones of *Allosaurus fragilis* from the late Jurassic of Wyoming contain pits considered to be the work of skin beetles that excavated pupal chambers after filling up on dried flesh.[127,128] The damage was extensive, appearing on at least 12% of the recovered skeleton.

Skin beetles were also considered to be the culprits that made pits in two bones of an Upper Cretaceous *Prosaurolophus* from Montana,[129] as well as circular-to-elliptical borings in Upper Jurassic sauropod and therapod bones at Dinosaur National Monument, Utah. These latter borings occurred on nearly 40% of the skeletal remains and were again interpreted to be pupal chambers. Scanning electron microscope examination of the bones revealed scratch marks attributed to the mandibles of the larvae. Apparently, the pits were formed some 4 to 9 months after the dinosaurs died.[127] It would appear that no object is too hard for the strong mandibles of skin beetles. These instruments, coupled with their strong desire to form pupation burrows, allow them to make tunnels in a number of dense materials, including horns, hoofs, mortar, stone work, and even lead.[130] A skin beetle in Burmese amber[131] could well have played a role in the recycling of dinosaurs. These ancient dermestids may have transported juvenile stages of tapeworm and nematode parasites to dinosaurs, just as they do to birds today.

### The Food Chain

The food chain in any ecosystem has two nutritional pathways based on what consumers eat. There are those that subsist on living organic material. This is the most well-known division, where energy is shunted through herbivores. The other division involves those organisms that consume detritus, dead or dying organic matter, where the energy moves through saprophagous organisms, also known as detritivores, decomposers, saprophages, and scavengers. In almost all terrestrial habitats the detrital food chain is dominant. Estimates indicate that 11% of the biota is comprised of saprophagous insects.[132]

The world remains green because herbivores do not consume all or even a major portion of living plants. The amount of floral production removed by plant predators varies between ecosystems, with an estimated low of 2% in a poplar forest, 10% in salt marshes, and between 33% and 66% in overgrazed dry savannahs. Overall, the total amount worldwide is probably around 10%, and about half of this is returned to the environment as feces.[66] So dropping leaves, withering grasses, and falling trees enter the detrital food web, along with animal waste and bodies, where they are broken down by saprophagous organisms. Insects share this nutritional bonanza with fungi, bacteria, and other invertebrates such as mites, snails, millipedes, nematodes, earthworms, and springtails.

Next to plants, detritivores process the largest amount of energy in food webs, much more than any herbivores past or present, including dinosaurs. In so doing, they return nutrients to the soil for plants to utilize. The majority of saprophages, including insects, become part of the food chain as prey for carnivores.

The importance of this predominant pathway in the food chain is frequently overlooked because it is not obvious. Large grazing animals and predators are conspicuous, but small organisms, especially insects, are primarily responsible for the everyday functioning of the biome. While some dinosaurs undoubtedly were carcass scavengers and others may have consumed herbivore feces like pigs and rabbits do now, they certainly were not a significant factor in the detrital food web. They were then, just as we are now, dependent on insects and others for processing detritus. Without saprophagous organisms functioning to remove waste, the world would suffocate under the accumulating masses of dead plant matter.

Perhaps this is why termites are such a successful group today. As probably the most significant insect detritivores on the planet, their more than 2,300 species, aided by internal symbionts, break down cellulose in wood and plant tissues. Their numbers are astonishing and individuals in a single nest can reach 20 million or more, while in some areas of the tropics, termite mounds occupy almost 30% of the soil surface.[66] Their efficiency in destroying human dwellings can be judged by the billions of dollars in damage they cause each year just in the United States.

So dinosaurs benefited directly from saprophagous insects because they were important nutritional items for their young and for smaller species, and also indirectly because they supplied a source of food for insectivorous mammals, reptiles, and birds that they fed upon. And most importantly, dinosaurs profited because saprophagous insects cleaned up and maintained their environment.

# 10.

## The Case for Entomophagy among Dinosaurs

*A particularly large beetle was rolling a dung ball away from one of many fecal piles dotting the over-grazed plain when it was suddenly pounced upon and gobbled down by a young ornithomimid. After consuming a few additional unlucky adults, the dinosaur started to dig around the perimeter of the drying excrement, searching for juicy dung beetle larvae, but they were buried too deeply and he left to try his luck elsewhere.*

*Further away at the perimeter of the dense forest, an immature pachycephalosaur was nibbling on cycad leaves when a sudden movement caught his eye. A large cockroach that had been resting in the dried leaves at the base of the thick stem had been exposed when the plant was disturbed. The youngster bent down and snapped up the delicacy, which represented a concentrated package of protein and fat needed by the growing animal. Deciding to probe around for more, the small juvenile pushed his head further into the fallen leaves at the base of the plant. As some earwigs crawled away from their now-divulged hiding place, they too were gobbled down. The dinosaur then proceeded to scrape away the overlying plant debris, uncovering the top layer of soil and thereby revealing some ground beetles that were added to the meal. When satisfied that no more insects were available at that spot, he then turned his attention back to the cycad leaves, since these and other plants would constitute the main part of this omnivore's diet.*

*Another diminutive dinosaur was inspecting a babbling brook for prey by systematically turning over partially submerged rocks and targeting mayfly and stonefly larvae. The disturbance caused eddies that suspended caddis and Dobson fly larvae, which had been resting underneath the stones. A hurried attempt was made to snatch them up before the current carried them downstream. Finally some of the blackfly larvae that crowded around the edge of the rocks just beneath the rushing water were pulled from their holdfasts and swallowed. Aquatic insects were plentiful and it was not difficult obtaining a good meal from the clear waters.*

In all probability, almost every dinosaur, even those considered vegetarians, were in actuality omnivores at some point in their lives, certainly when they were in the rapid growth stages and possibly also during periods of egg production. You may question how we arrived at that conclusion, but even today characterizing an animal as an herbivore, omnivore, or carnivore is an almost impossible task, so we assume the same held true for dinosaurs. Few vertebrates are truly one or the other. For example, mammals begin their lives dependent on a food source rich in protein, mother's milk. Fruit- and nectar-eating birds feed bugs to their young, and because many insectivorous birds take fruits and seeds as well, most birds would be considered omnivores.[133] So both mammals and birds start life eating animal proteins, and although some go on to become herbivores as adults, overall they could be classified as omnivores. Since some paleontologists consider birds living dinosaurs, the principle of behavioral fixity dictates that omnivory was common among dinosaurs as well.[30,2] Also, scientists are discovering that a large number of animals traditionally considered herbivores or carnivores will cross over when given the opportunity, especially during periods of plenty or scarcity of one type of food.

Some animals consume the most unexpected things. For instance, there are flesh-eating deer, hippos, chimpanzees, and hamsters, and fruit-eating wolves, bears, badgers, and mongooses. The list becomes even more extensive when you look at who in the animal world ingests insects at some time or other. In fact a much easier way of looking at it would be to identify only those that don't! After all, eating arthropods makes a lot of sense. They are the most abundant animals around and represent a readily available source of protein. The same range in omnivory seen in present-day animals would probably also apply to dinosaurs. Probably most herbivorous dinosaurs, and certainly all of the carnivores, actively searched for and consumed insects during their growing years. Insects could be regarded as the ideal convenience food of the Cretaceous.

Vertebrates aren't the only ones whose eating preferences don't fit nicely into a definite category. In almost all insect families, there is some crossover from herbivory to carnivory or vice versa. Sometimes one type of feeding behavior is characteristic of the larvae and another of the adults. For example, mosquito larvae ingest plant debris and microorganisms, while adult females are mostly bloodsuckers. Almost all bloodsucking flies also imbibe sugar from plants or nectar-producing insects for long-term survival. Grasshoppers and their kin often stop to partake of dead or dying insects, and cockroaches are well known for including both animal and plant matter in their meals. Even primitive springtails add nematodes to their diet of pollen and spores. Many plant bugs will lunch on other invertebrates encountered on leaves. And larvae of many aquatic insects, like caddis flies, subsist on both plant and animal matter they scrounge from the bottoms of ponds and streams.

### Passive Feeding

As herbivorous dinosaurs munched away on ferns, cycads, and conifers, they passively consumed hundreds, even thousands, of insects in their food (color plates 5B, 12C). Could the passive ingestion of insects by megaherbivores be regarded as an important food additive? Initially, we assume that a sauropod weighing 80 tons and eating thousands of pounds of plants per day was hardly affected by say, a few pounds or so of accidentally consumed bugs. But consider that aside from the obvious proteins, fats, and fiber, insects contain a number of vitamins and minerals that might be regarded as beneficial.[134] Proteins, vitamins, minerals, or even chitin in incidentally swallowed insects may possibly have been required for the maintenance of vital intestinal symbionts of some dinosaurs. Lack of insects in their diets might have meant impeded digestion, poor health, or a greater susceptibility to disease. Perhaps there were components in insects required by dinosaurs for hormonal production.

On the other hand, we know that many insects sequester toxins from plants today, and in the Cretaceous, they may have killed or sickened dinosaurs that unknowingly ate them. The effect of passive ingestion of insects on dinosaurs is purely speculative at this point, but remember that any action, no mater how seemingly insignificant, is not without some consequences. Even if the passive consumption of insects by dinosaurs had no discernable effects on the eaters, it certainly had one on the eaten, possibly even reducing populations of some phytophagous insects.

Although we have, in the United States, stringently controlled guidelines for food purity, humans unknowingly ingest numerous insects in prepared products. A spaghetti dinner, for example, could contain over seven hundred bits and pieces of arthropod parts and still be considered acceptable under USDA standards. This is still quite a bit less than mankind consumes in other parts of the world and considerably less than the intake by herbivores past and present.

*Active Searching*

There can be little doubt that insects represented an important component of the food chain in the Cretaceous. As primary consumers, insects certainly served as an energy source for many small animals, which in turn were consumed by larger animals, and so on. Insects provided an easily available source of much-needed protein for precocial dinosaur hatchlings that received no parental feeding.

We believe that most, if not all, dinosaur young fed on insects. How long this dependency on them remained cannot be determined. Many large present-day mammals, including humans, consume insects as an occasional or even regular part of their diet. Some of these animals can hardly be considered diminutive, with bears topping out around 800 pounds and pigs at 500-plus pounds. It follows that dinosaurs of similar proportions also incorporated insects into their diet regardless of whether they were carnivores or herbivores. The importance of insects and other invertebrates in any food web cannot be over emphasized. They are frequently forgotten or dismissed because of their size, but they represent more than half of the biota and without them, much of the food chain would collapse.

Active searching for insects by dinosaurs presumably occurred in multiple habitats. After storms, some probably snapped up stranded leaf beetles and moth larvae that had been defoliating various plants. We can easily imagine scale insects on twigs being lapped up with long, mobile tongues, and dinosaur young might have even relished the sweet exudates produced by these tiny insects, just as children delight in sugary sweets.

During the entire year, adolescents were on constant lookout for large arthropods. Defoliating insects on ferns, conifers, and angiosperms served as a snack, but hefty wingless walking sticks, jumbo cockroaches, and giant preying mantids provided better meals. Certainly the most abundant of these were cockroaches (color plate 6D) that occurred under rocks, in debris, around piles of dung, on and under dead animals, and scrambling over tree trunks. Their associations with dead animals and dung also made them important vectors of dinosaur parasites such as protozoans and stomach worms.[135] Various types of larger orthopterans were also available for consumption and undoubtedly provided excellent meals. These included crickets, katydids, monkey grasshoppers, elcanids, wetas, and mountain crickets or haglids. The latter feed high in the trees at night and might present a challenge to all but arboreal predators. One of the few remaining representatives of this family subsists on staminate cones of conifers.[136] The males sing to locate mates, but in the Cretaceous their love songs may have attracted hungry dinosaurs.

Crickets and grasshoppers (color plates 4A, 4B, 8A) made choice dietary morsels throughout the Cretaceous, either plucked from their feeding places on the foliage of conifers or collected from the surface of ponds when the wind blew them off course. Some of the Cretaceous grasshopper-like elcanids[50] probably built up high populations at certain seasons, stripping the leaves from plants (color plate 8A). The chances are good that they formed swarms, just as locusts do today. When populations peaked, their massive hordes were sought out by a variety of dinosaurs, as well as by other vertebrate and invertebrate predators. Searching in the soil for mole crickets and around the bases of tree ferns, cycads, palms, and bamboos for wetas and king crickets presented a more challenging job for insectivorous dinosaurs, but many birds and lizards currently are successful at it. And dinosaurs that were accustomed to opening plant stems could have dined on large, juicy beetle and moth larvae.

Some ornithomimids and even troodons possibly collected aquatic insect larvae by turning over rocks in the streams or browsing the detritus-littered bottoms of ponds. Selected tidbits like mayfly, stonefly, and dragonfly nymphs, especially when crawling out of their aquatic homes to enter the terrestrial world, made excellent meals. Just as humans collect and eat caddis fly larvae,[137] the dinosaurs could have done the same millions of years ago (color plate 7C).

Ponds supplied snacks such as giant water bugs and large water beetles, although even the smaller backswimmers and water boatmen would qualify if enough were collected. Depending on the speed and agility of the individual hunters, dragonflies and damselflies plucked from the tips of reeds or possibly knocked to the ground from the air made a good mouthful. At certain times of the year, large swarms of midges and lake flies (chaoborids) amassed over lakes and ponds, mating and laying eggs, and within a few days their bodies piled up inches deep along the shores. Even if the dinosaurs had to venture onto the mud flats to scoop them up, the huge numbers of these tiny insects surely made it worth the effort. Along the borders of the marsh grew rushes with leafy galls containing populations of psyllids. These particular rushes might have been selected as food items by adolescent saurichischians after they had relished the tubers and corms of sedges infested with moth and weevil larvae.

While trudging from one feeding site to another, there was always the chance of encountering a large spider or scorpion. These arachnids would have been a delicacy, just as they are in some human societies.[137,138] Whether the scorpion's sting discouraged some of the dinosaurs is difficult to say, but extant birds and mammals appear to have no problem eating them.

At certain seasons, many insects aggregate for mating or dispersal and are easily captured. So June beetles periodically swarming around dying gingko trees, stink bugs amassing on conifer foliage, and even pygmy grasshoppers (color plate 7B) congregating within some of the larger tree buttresses were presumably welcomed, perhaps even anticipated, and then devoured *en mass* by dinosaurs.

The periodic emergences of cicadas certainly attracted the attention of some dinosaurs as the brownish nymphs slowly crawled up tree trunks or flutter-dried their wings after shedding their drab outer coat. The presence of these fairly large insects without a doubt became known to all after they began their deafening screeches. At this time, it is likely that agile dinosaurs plucked them, still shrieking, from the branches. The calories that these insects had accumulated after years of feeding on plant roots provided predators with a good energy source.

Insects that breed relative quickly on plant foliage can build up populations large enough to be utilized as food. Planthoppers comprise one of these groups (color plates 2A, 2B). Normally small, often beautifully colored and curiously shaped, these insects feed on plant juices and together with spittlebugs and leafhoppers presumably added some variety to the diet of small insectivorous dinosaurs.

The sudden appearance of thousands of shimmering specks in the sky has signaled the emergence of millions of winged termites throughout time. Humans in various cultures herald the arrival of these insects, and they certainly attracted the attention of dinosaurs. Termites provide a good meal easily gathered in a short amount of time. My experience with eating termites came when I was working in Burkina Faso on a project funded by the World Health Organization some years ago. A heavy downpour had thundered on the tin roof of our compound for about ten minutes before passing. When I went outdoors to feel the refreshing cool air that followed the storm, I was astonished to find thousands of large, winged termites rising up from what appeared to be every square foot of soil as far as I could see. The cook was already out carrying a pail of water in one hand and enthusiastically grabbing at the fluttering termites with the other. Upon securing one, a quick movement of the wrist plunged the helpless captive into the water. After collecting several hundred, the cook prepared them by stripping the wings from their bodies and throwing the still active creatures into a pan of boiling water. Then everybody in the camp sat around and dipped into the communal pot, pulling the floating limp bodies from the oily water and popping them into their mouths. After contemplating a greasy one in my hand for a while and wondering if their intestinal symbionts formed heat-resistant stages, I decided that I really preferred them fried, but the others obviously relished them.

Fossil evidence confirms that termites existed throughout the Cretaceous (color plate 7A). Those opportunistic feeding binges they afforded dinosaurs when emerging only lasted a few hours and were periodic events, occurring just several times during the year. How did dinosaurs take advantage of these social insects at other times? Many types of troodons probably used their large curved claws to break through the hardened walls of termite mounds. Their narrow snouts made them well suited to enter the damaged mounds and snap up the tender workers. With some additional effort, they might have been able to reach the royal chamber and find the choicest item of all, a large queen so filled with protein-rich eggs that she was unable to move. Of course, the soldier termites were not going to stand by idly while their home was invaded, and they would have rushed to sink their huge mandibles into the lips of the marauders. Those predators that didn't have the strength to break into a termitarium could have simply waited by the openings for the workers to begin their daily tasks and then gobbled them up.

We can assume that dinosaurs also ate ants, which were present at least by the mid-Cretaceous (color plates 6A, 6B, 6C). While their queens are not much larger than workers, they contain a concentrated sac of proteins, and it is plausible they were eaten as they emerged from the old nest or while searching for new nesting sites. Some of the ants back then might have made nests inside living or fallen partially decayed trees. Those dromaeosaurs that preferred ant eggs, larvae, or pupae only had to destroy the nests to reach the brood. If some Cretaceous ants lived in climates with a relatively long dry period, additional delicacies known as replete workers lurked inside the nest to tantalize the predators. The swollen bodies of these workers served as receptacles of sugary material collected by the foragers. Today, Australian aborigines,[139] as well as a host of birds and mammals, spend hours digging up these honey-pot ants from their subterranean homes.

While we as yet have no evidence that Cretaceous ants used this method of storage, aphids and scales were there to produce honeydew. Aside from a few flowers, most sugars likely came from excretions of scale insects and aphids, insect groups that were abundant throughout the Cretaceous. Many of these small sugar factories turn plant juices into delectable metabolic compounds that are a preferred food of many extant ants. And they are so important to the ant colony that when new queens leave their homes in search of fresh adobes, they will often take along one of the sugar-secreting scale insects to insure a supply for their progeny. So did dinosaurs have a taste for sweets, like birds and lizards do today,[56] and if so, would they have enjoyed a little sugar along with their protein meal of insects?

The raiding of ant nests wasn't without inherent problems since Cretaceous worker ants were equipped with large powerful stingers that probably released painful toxins, and they never hesitated to use them against anything threatening their homes, including dinosaurs. So marauders would have had to pause in their quest from time to time to wipe them away. Perhaps those irritations acted as a deterrent and prompted some to quickly decide that putting up with such insults wasn't worth the benefits of a few juicy ant larvae. But if a few stings did not matter that much, insectivorous dinosaurs with a preference for ants probably just stood on a mound and consumed the defenders that rushed from the nest and swarmed onto their feet and legs. In cases where the ants lived in trees, the workers were undoubtedly snatched directly from the leaves, branches, and bark.

What about bees? When we think of bees, we think of social bees that store the honey we find so delectable. However, the great majority of bee species are solitary, like *Melittosphex*, and only store a mixture of pollen and nectar for their larvae, much too little to interest most vertebrates. While records of Cretaceous honey-producing social bees have not yet been confirmed, primitive stingless social bees similar to those found in the tropics today may have appeared in the Late Cretaceous when angiosperms were more prominent. If these little bees were around, their comings and goings very likely were noticed, and dinosaurs may have even followed workers that were collecting pollen back to their nests or detected the small bees entering and leaving the tiny wax-coated holes on the trunks of dead standing trees or in the ground. While the nests located at the tops of the trees were out of reach, those at the base of the trunk or in the soil were available for raiding. Some canny individuals possibly detected the faint hum that these bees produce and only had to dig to find the honey-filled cells in ground nests.

Upon uncovering the nurseries, dinosaurs would have been rewarded with a variety of edible treats, including large wax cells containing pollen, nectar, and honey, as well as smaller ones with eggs, larvae, and pupae. This mixture of wax, sugary material, and insects was probably quickly gobbled up. While these small bees lacked functional stingers, they still swarmed out of their nests to attack the intruders. Their main means of defense was their mandibles, and although even smaller than a housefly, they crawled over the faces of the looters, entering their ear openings, nostrils, and mouths. Aside from biting into these membranous areas, some of the workers could have smeared a wax-like greenish material over the eyes of the assailants, obscuring their vision and causing them to stop and wipe their eyes. Since their liquid, dark brown honey is often stored together with older supplies that have become fermented and sticky,[137] the raiding dinosaurs must have emitted a strong stench when they finished their meal.

Aside from looking for ant, termite, or bee nests in old trunks, dinosaurs certainly ripped open rotting logs to collect the large juicy larvae of lucanid, oedomerid and longhorn beetles, just as we see crows use their beaks to break open logs and dine on such resources today. And while they were examining exposed logs, some of the smaller beetle larvae, like those of the click beetles, tumbling flower beetles (color plate 5D), passalid beetles, and flat-headed borers undoubtedly were used as a food resource.

All dinosaurs would have eaten insects at one point or another during their development, especially when they were growing. We can assume that those dinosaurs that were basically herbivores as adults passively consumed many insects along with their plant food, while the insectivores and even some larger carnivores persistently fed on them throughout their lives.

# 11.

## Gorging on Dinosaurs

FRESH VERTEBRATE BLOOD is not exactly everyone's ideal meal. But for a few animals such as leeches, vampire bats, and hematophagous insects, it represents haute cuisine. And even some humans, like the Masai of Kenya, are well known for surviving on a mixture of milk and blood drawn from cows and goats. But by far, the largest group of animals to develop this proclivity is bloodsucking insects, and they have truly perfected this habit because their very survival has depended on this sanguinary diet since the Cretaceous.

Even when we slather ourselves with repellants, bloodsucking insects can make our lives miserable. How could something so small cause so much discomfort? Not only do they annoy us by buzzing around our heads and crawling over our faces, but they inflict wounds that are downright painful and leave swellings that can itch for days. Armed and dangerous, we call them biting insects; however, they really don't bite. While we chomp by placing our teeth around an object and then bringing our jaws together, biting in bloodthirsty insects is much more complicated, and if this sophisticated feeding system didn't result in so much human suffering, we might admire it as a marvel of invention.

They begin with the arduous task of piercing or cutting away a small portion of skin, adding saliva that contains anticoagulant factors, and then finally sucking up the blood. The job is performed with minute needle-like appendages called stylets, and every insect group has their own particular modifications. Most have narrow, elongate mandibles lined with serrated edges much like steak knives. To assist, another pair of slender, serrated structures called maxillae is often employed. Positioned between this set of cutting instruments is a trough-shaped structure, the hypopharynx, containing a channel through which salivary secretions pass down into the wound. Protecting the mandibles and maxillae from above is the labrum, the underside of which contains a food channel that carries blood from the victim up into the insect. These precision cutting tools resemble delicately designed surgeon's implements.

Each type of bloodsucking insect has its own unique but equally effective method of penetrating vertebrate skin (fig. 22). Mosquitoes are capillary feeders. They first insert their stylets beneath the skin and search for a blood vessel. Once found, the wall of the capillary is punctured, the mouthparts enter, and blood is sucked up directly into the alimentary tract through the food channel. There is usually no fluid left on the surface of the skin when the operation is finished. Blackflies, biting midges, sand flies, and horseflies employ the less-refined pool-feeding method. A hole is cut in the skin, severing several capillaries. Blood is drawn up as it collects in the open wound. Upon completion, there is usually some residue left on the surface of the skin, which often attracts other insects.

The blood meal consists not only of plasma and cells, but any microorganisms that might be developing inside these cells or floating free in the plasma. These pathogens are then passed on to the next victim with the saliva, explaining how biting insects acquire and spread infectious agents.

The main function of the blood meal, which is approximately 20% protein, is to supply nutrients for egg production. It also provides some nourishment for the adults, and mosquitoes have been kept alive for over 80 days on some 30 blood meals.[140] Aside from blood, biting flies generally require sugar supplements for long-term survival.[141,142]

These are obtained from various plants (nectar, nectary glands, fruits, plant juices) as well as from insects in the form of honeydew, and also appear to be important for the development of some pathogens they might carry. Female sand flies obtain sup-

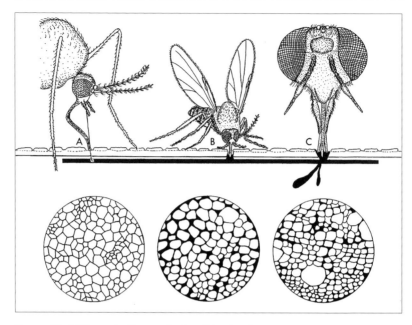

Figure 22. Different feeding methods of bloodsucking insects, and dinosaur scales. A. Capillary or tube feeding by mosquitoes. B. Micro-pool feeding by biting midges. C. Macro-pool feeding by tabanids. Below are scale patterns of dinosaurs and a chameleon. Left, a sauropod. Note the angular edges of the scales and their close fit. Middle, a hadrosaur (*Corythosaurus*). Here the scales have rounded edges with more exposed skin between them. Right, a mountain chameleon (*Chamaelo montium*). The scale pattern and scale dimensions are closer to that of the hadrosaur. All drawn to the same scale. Dark areas represent exposed skin surrounding the scales.

plementary plant juices by "biting" stems in the same manner they penetrate skin.[143]

How do hematophagous insects locate and track their victims? Detecting a host can be accomplished visually, but odor also plays an important role, and many compounds in vertebrate skin serve as attractants, especially carbon dioxide. Even small amounts of this gas, such as from an animal's breath, can grab their attention. Heat also is a lure and many biters are able to discern the smallest temperature gradients in the atmosphere.[142]

Some bloodsuckers are generalists and attack mammals, birds, reptiles, and amphibians. This definitely favors their survival since if one host is unavailable, another is there to replace it. Others are more finicky and target only a few selected animals. How many blood meals does a biting fly take during a lifetime? Most imbibe at least twice, which is just enough to supply nourishment for two egg clusters and also the basic number needed for pathogens to be both acquired and transmitted.[142]

The various types of biting arthropods that we believe fed on and transmitted pathogens to Cretaceous dinosaurs are discussed in the following pages. Some, which probably gorged on dinosaurs but are not discussed, include snipe flies (Rhagionidae) and athericid flies (Athericidae), which attack mammals today and existed throughout the Cretaceous. Their modern counterparts are not known to vector any microorganisms. Also extinct Cretaceous chironomids, tanyderids, corethrellids, and mecopterans that had biting mouthparts were also capable of feasting on dinosaurs, but whether they were carrying pathogens is unknown (color plates 10, 11A, 11B, 11C).

### Piercing Dinosaur Skin

When one considers the immense size and strength of many dinosaurs like sauropods and tyrannosaurs, it might appear that tiny biting insects would be up against insurmountable odds. Surely such animals had thick, impenetrable skin. But the reality is that except for the back and sides of plated stegosaurs and armored ankylosaurs, dinosaur skin was surprisingly thin and reptilian, in fact very similar to that found on present day chameleons and Gila monsters (fig. 22). How amazing that creatures 40 or even 100 feet long and weighing many tons had the same size and type of scales as these small lizards!

The dry skin of reptiles is covered with scales, which are localized thickenings of the horny material keratin situated in the stratum corneum of the epidermis. They usually form a continuous layer and vary in size, shape, and texture. Imbricate scales overlap but scales of some lizards are juxtaposed to one another with exposed bare areas of unprotected skin between them. Scales can be small and polygonal, large and plate-like, and smooth or possess various ridges or keels. Size may be directly related to flexibility, with small ones located on mobile areas and larger ones on immobile areas such as the top of the head or back. Certain types of reptiles have tuberculate scales that are modified into wart-like raised extensions or tubercles. Scales are periodically shed with the skin as an entire sheet or in pieces of various sizes. Sometimes individual scales are discarded separately. This process occurs more frequently in growing stages, and the intervals between shedding vary between species. Whether dinosaurs molted periodically is unknown, but given their skin covering, highly likely.

Only a very few fossils of dinosaur skin have been found, and over half of these were from hadrosaurs, which were covered with tuberculate scales that fell into three categories.[145] The first consisted of uniformly small, ground tubercles with no definite arrangement. The second type was flattened, polygonal-shaped pavement tubercles ranging from 0.19 to nearly 0.25 of an inch in diameter, with smooth surfaces and mostly uniform margins. The third type resembled very large radially sculptured, limpet-like cones reaching 1.26 to 1.50 inches in diameter and 0.30 of an inch in height. One hadrosaur from the Late Cretaceous of Canada, *Edmontosaurus,* had extremely thin skin covered with ground and pavement tubercles.[145] Another duckbill, *Anatosaurus,* had ground and pavement types with larger scales on the sides, back, and tail and smaller ones on the shoulders and inner thighs.[148] Finally, the skin of *Corythosaurus* was composed of flat, polygonal tuberculate scales.[147] Many hadrosaurs had a skin fold or frill adorned with variously sized tubercles extending along their backbones.

Examples of ceratopsian skin have also been recovered. The integument of the horned dinosaur *Chasmosaurus,* also a Late Cretaceous specimen, was composed of polygonal, non-overlapping scales.[149] Another ceratopsian, *Centrosaurus,* had a similar integumental pattern with large tuberculate plates surrounded by smaller ones. Relatively thin skin with tuberculate scales was noted on two specimens of the ornithopod dinosaur *Iguanodon* from the Early Cretaceous.[146,150] In one, the skin was covered with scales ranging from 0.14 to 0.23 inches in diameter, and in certain areas they ran together and the skin appeared almost smooth. The outer surface of the other was composed of very thin skin dotted with small convex tubercles varying from 0.13 to 0.24 inches in diameter. In some areas, these were larger and flatter, while in others, the tubercles appeared to coalesce and were almost imperceptible.[150]

Sauropods such as *Camarasaurus* had epidermis with larger hexagonal, non-overlapping convex plates ranging from 0.43 to 1.10 inches in diameter. The plates in this tuberculate integument became smaller and rounded towards the armpits. The skin of another sauropod was also relatively thin and the scales formed a rosette pattern. These were 0.80 to 1.20 inches in width and covered with minute papillae. On portions of the skin were dermal spines similar to those on the backs of iguanas and along the tails of crocodilians.[151]

Pebbly skin impressions of the tyrannosaurs *Gorgosaurus* and *Daspletosaurus* show that the common ground plan of tuberculate scales ran through the entire dinosaur line. The armored dinosaurs, ankylosaurs and nodosaurs, along with the titanosaur sauropods, bore dorsal and/or lateral bony plates, also known as body ossicles, osteoderms, or scutes, similar to those found on turtles.

What was the skin surface in feathered dinosaurs? The genera *Sinosauropteryx, Protoarchaeopteryx, Caudipteryx,* and *Beipiaosaurus* from the Cretaceous of China are all described as having feathers on part or all portions of their bodies, including their tails, and some even consider these fossils as birds. While a few paleontologists believe that the protrusions covering these fossils are a kind of quill-like extension of the scales, others feel that there is no doubt about them being true feathers.[152] Although the actual integument of these dinosaurs was not preserved, in present day birds, feathers arise from discrete areas surrounded by membranous skin. All groups of biting arthropods discussed in this book feed on birds, so the presence of feathers on dinosaurs would not deter the bloodsuckers.

So where would these arthropods feed? At first glance, it would appear that obtaining blood through the scaly skin of dinosaurs was an impossible feat. However, there were several ways available. If the scales were thin, arthropods had no problem penetrating directly through them. When too thick, the biter might have utilized minute openings in the scales similar to those occurring on some extant reptiles.[144] While only a few microns wide, the holes in all likelihood were enlarged with the cutting edges of the mandibles or maxillae. However, if these openings did not occur on dinosaur scales, arthropods were capable of feeding on the softer, non-keratinized skin between the scales. The thin integument exposed around the tubercles was the Achilles' heel of dinosaurs and definitely offered no protection against bloodsucking arthropods. These biters could simply have fed around the margins of the tubercles anywhere on dinosaurs and did not need to search out specific, less-keratinized locations such as around the mouth, eyes, ears, or nasal openings, although they also were prime sites. The loose skin lobes on the forelimbs of some hadrosaurs also offered choice feeding areas. Even the dermal plates of *Stegosaurus*, which were lined with blood vessels on their outer surfaces, undoubtedly provided feeding places.[30] And if dinosaurs periodically shed their skin, they became particularly vulnerable to biting arthropods at those times.

The similarity in size, shape, and patterns between the scales of dinosaurs and chameleons is striking[154] (fig. 22). Since chameleons are bitten by mosquitoes, ticks, and sand flies, all of which introduce pathogens into these colorful lizards, the dinosaurs certainly were sitting ducks for bloodsuckers too. They probably had few means of avoiding the pests, unless the small forelimbs on some could have been employed to shoo away flies or scrape off ticks. Some relief may have been found in water, and perhaps the large sauropods, and even hadrosaurs, submerged their under parts as much as possible for protection from the voracious biters.

The wide host range of most present-day bloodsucking arthropods indicates that it really wasn't significant if dinosaurs were warm- or cold-blooded[155] or if they were covered with tuberculate scales or feathers. These formidable micro-predators still would have been able to locate feeding sites on Cretaceous dinosaurs just as they are able to use birds, mammals, reptiles, and amphibians as a smorgasbord of protein delights today.

# 12.

## Biting Midges

*An industrious hadrosaur stripped foliage from a kauri tree sapling growing at the border of a sodden meadow. Metallic-colored dragonflies greeted the day by darting across the tops of ferns and horsetails and catching small insects on the wing. Clouds of silvery-winged mayflies fluttered up from the dew-covered vegetation. The dinosaur munched away at overhanging branches, consuming myriads of aphids and scale insects that covered the leaves. The commotion caused small weevils and xyelid sawflies to drop to the ground, where they unfortunately encountered beetle-shaped scavenging cockroaches. Primitive katydids and walking sticks escaped the encroaching teeth by fleeing to quieter areas of the tree.*

*This idyllic scene was soon disrupted by a group of hungry biting midges that had been resting on some nearby reeds. Rising into the air like a puff of dark smoke, they headed with bloodthirsty determination straight for the pebbly skin of the feeding animal. The jabs of sharp pain felt by the browsing hadrosaur several moments after they alighted were probably ignored. The voracious insects took about two to three minutes to fill their stomachs with blood, and then rested briefly before returning to the reeds. The females had obtained their goal, a source of protein, and would soon lay their eggs in moist soil at the edge of a swamp. Their bites were not innocuous because their victim may well have been infected by microscopic parasites that would take its life. Only time would tell.*

Biting midges (ceratopogonids) were just one of many insects that fed on vertebrate blood in the late Mesozoic. From amber, we know they shared this habit with sand flies and corethrellid flies,[338] as well as other groups found in different types of fossil deposits (color plate 11C). Some people may be familiar with biting midges as the minute "no-see-ums" or "punkies" that deliver painful bites. You may not notice them coming or going—but you definitely know when they begin to feed! These ancient flies were quite common and widely distributed throughout the Cretaceous.[13,156] Their evolutionary success is due to their broad host range, which includes taking blood from insects, fish, amphibians, reptiles, birds, and mammals. But they also have adapted to many different habitats that can support their larvae, from aquatic (lakes, streams, tree holes) to moist (decaying plant material, dung) and even semiarid (sand).[157,158] We can say with a high degree of confidence that the developmental sites of Cretaceous biting midges probably included dinosaur dung.

With such a wide range of available food sources, how do we know if a fossil biting midge could have fed on dinosaurs? Fortunately there are some morphological features that directly tie them to a specific feeding behavior. Those that feed on vertebrates (color plates 9A, 9B) tend to have finely serrated mandibles, retrose (downward pointing) teeth on their maxillae, and small tarsal claws, while invertebrate feeders generally have coarsely serrated mandibles, maxillae without retrose teeth, and large tarsal claws. Additional characters also provide clues as to what type of vertebrates are attacked, such as whether they are warm- or cold-blooded, birds or mammals, and even their relative size. An analysis of these characters on Cretaceous biting midges indicated that several genera and species dined on large vertebrates, which certainly included dinosaurs.[51]

Those that relished dinosaur blood could have been related to biting midges that currently feed on lizards[160,161] or attack turtles and iguanas. A leptoconopid midge searches out several species of California sand-dune lizards on cool early mornings.[160] Once they insert their mouthparts and start engorging, they are almost impossible to remove, even when the lizard crawls through the sand. These biters belong to an ancient lineage that probably attacked dinosaurs.

The onslaughts of ceratopogonids on sea turtles commence as soon as the reptiles arrive on land and continue until they reenter the sea.[162,163] In fact, these cunning insects seem to anticipate their host's arrival! They gather in groups of seven or eight and over a hundred individuals will dine on a single turtle during the forty-minute period it is out of the water. This association between biting midges and sea turtles along Central American beaches is probably quite ancient because the insect's distribution coincides with the nesting sites of the leatherback turtle. There appears to be a decided preference for these marine reptiles, since as long as they are present, humans and dogs are not bitten.

Did dinosaurs have allergic responses to biting midges? Many humans experience several types of reactions to these biters. Aside from the initial searing pain, there can be a response to their saliva, expressed as a tender swelling around the wound. In sensitized individuals, itching blisters can leak fluid for two to three days and scratching can cause bacterial and fungal infections, which without treatment could become life-threatening.[157]

Since biting midges are so minuscule (color plates 9A, 9B), what part of the dinosaur would they attack? These insects search for a highly vascularized area of the epidermis. In elephants, this region occurs behind the ears.[158] The skin between the tubercles would be the natural point of attack, and even if some dinosaurs were covered with feathers, the surface between the quill insertions would provide adequate feeding sites. Ceratopogonids are pool feeders[157] and all they require is an area of thin skin with some capillaries close to the surface. As soon as blood fills their small trough, they begin dining. Some may have favored the area around dinosaurs' eyes like the site chosen by the present-day human biter, *Austroconops*.[51,166]

If one of the no-see-ums that gorged on a dinosaur contained stages of Cretaceous malarial parasites, such as *Paleohaemoproteus* (figs. 23, 38),[167] the microorganisms could have been transferred into the vertebrate's bloodstream. At first, the dinosaur may have felt no ill effects. However after the protozoa began to

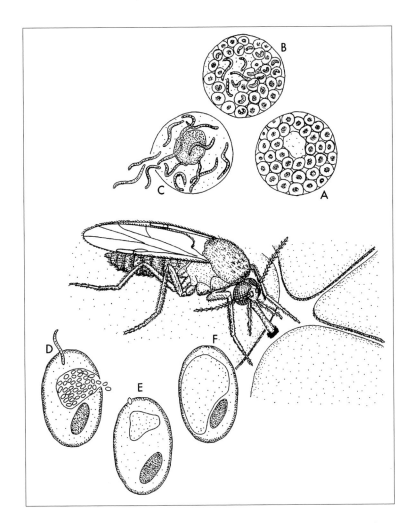

FIGURE 23. A Cretaceous biting midge (*Protoculicoides* sp.) carrying a malarial parasite (*Paleohaemoproteus burmacis*)[167] is feeding on the exposed skin between the scales of a dinosaur. Top, stages of the parasite that occur in the biting midge: A. Spheroid bodies destined to become sporozoites developing in an oocyst attached to the gut wall. B. Immature sporozoites developing in the oocyst. C. Mature elongate sporozoites leaving the oocyst and migrating through the vector's body to the salivary glands. Bottom, stages of the parasite that occur in the dinosaur: D. Elongate sporozoites enter cells of internal organs and produce small, spherical asexual bodies called merozoites. E. One merozoite enters a blood cell while another has already begun development into a gametocyst. F. A mature gametocyst nearly fills a blood cell. This is the stage acquired by the biting midge while feeding. Sexual reproduction of the parasite occurs in the insect's gut, and the product of the union (ookinete) enters the gut wall and develops into an oocyst. Continue the parasite cycle by going back to A. Dark oval objects in the vertebrate cells are nuclei. Not drawn to scale.

multiply, she could have suffered anemia, weight loss, and possibly death. It would not have mattered whether she was warm- or cold-blooded since this type of malaria develops in a wide range of birds and reptiles today[157,168–170] and vectors were present throughout the Cretaceous.[156,164]

Another punkie may have been infected with vertebrate pathogenic viruses. Numerous viruses in the family Reoviridae are transmitted to mammals and birds today by ceratopogonids.[157] One that causes bluetongue disease (an *Orbivirus*) is highly lethal in ruminants and has caused the deaths of over a million sheep in Europe since 1998.[165] Other ceratopogonid-transmitted reoviruses cause lethal diseases in horses and infect wallabies and kangaroos.[171] Burmese amber biting midges were infected with reoviruses (color plates 9C, 9D), so these pathogens were certainly present by the mid-Cretaceous.[172]

These flies are also capable of vectoring other kinds of viruses to birds and mammals. One, a rhabdovirus, causes bovine ephemeral fever and several related diseases of cattle in Africa, Asia, and Australia.[164] There is even a report incriminating a biting midge in transmitting the Charleville rhabdovirus to reptiles in Australia,[171] and no-see-ums also carry viruses in the family Bunyaviridae to rabbits and ungulates.[157]

From the above accounts, it is obvious that these minute insects are important vectors of arboviruses (arthropod-borne), and since at least one type of reovirus is known from the mid-Cretaceous, others could have been infecting dinosaurs.

Biting midges also disseminate filarial nematodes similar to one that causes a widespread skin condition in monkeys, the great apes, and humans in Africa and South America.[157] The roundworms responsible occur in the subcutaneous tissues and body cavity. Human infections were probably acquired by midges feeding on both men and primates because the nematodes in these hosts are closely related.[173] Ceratopogonids also transmit filarial nematodes (*Splendidofilaria*) to magpies and quail in North America. In both cases, the parasites lodge in the wall of the bird's aorta.[173]

Since biting midges are cosmopolitan, occur in many different habitats, and feed on all groups of vertebrates, their potential role in disseminating diseases in the Cretaceous could have been quite significant. While they are small and have a limited flight range, winds can carry infected individuals over long distances, possibly as much as a hundred miles in a single night.[165] In the case of African horsesickness disease, winds carried infected midges from Africa to the Middle East, Cyprus, and Turkey.[157]

Some ceratopogonids would have sought their victims during the day, especially in the early morning or late afternoon just before sunset. Others would have fed at twilight and/or at night. The biting cycles were probably linked to the activity of their preferred hosts. Since these insects are active in subtropical and tropical areas throughout the year, there would have been little respite for the dinosaurs.

# 13.

## Sand Flies

On a cloudy, damp afternoon, a hungry gecko was cautiously crawling over the trunk of an araucarian tree, looking for insects but at the same time watchful for enemies. In one area a portion of the bark was missing and several beetles had collected, all of which provided a potential meal. The lizard moved to grab one but they all darted away, and in the end, the reptile chose a small weevil for its next meal, one that was crawling along the tree and lifting its elbowed antennae to detect scents of the preferred host plant. In an instant, the gecko lunged forward and caught the insect, but lost his footing in the process and fell into a sticky pool of resin. After struggling for several minutes, the animal became hopelessly immersed and died even before his partially swallowed prey showed any signs of digestion.

Hungry sand flies that had been hopefully following the luckless gecko now were forced to look for other prey. These miniscule, gangly insects with their long, slender legs, hirsute wings, and extended, flexible antennae appeared to be quite fragile but in actuality were astonishingly resilient. They didn't have to search long because a group of sauropods had entered the forest and began feeding. Like ghosts, the sand flies flitted through the shadowy trees and settled on the dinosaurs. After a few hops on the pebbly backs of the victims to find feeding sites, the flies became immobile as they began to gouge away miniscule portions of dinosaur skin. While drinking the red fluid oozing from the wounds, the females were joined by some males of the species, which, while lacking the necessary cutting tools, were only too grateful to partake of blood exposed by the females. After engorging, the fe-

*males were too heavy to fly and rested on the victims for several minutes, just long enough to excrete the liquid portion of the blood. They remained still while the expanding droplets of plasma issued from the tips of their abdomens and fell onto the skin of their hosts. With their stomachs now filled with concentrated blood cells, they laboriously flew to the overhanging branches of an araucarian tree, where they settled down to digest their meals. All, that is, except for one satiated fly, which by some miscalculation landed directly on a trickle of resin and would soon be inundated by another resin flow. In a few days, the surviving females would deposit their eggs and then be ready for yet another meal of blood.*

Bloodletting is a medical practice used by humans for well over two thousand years. The procedure, known as phlebotomy, involved puncturing one of the larger veins and draining blood into a container. This process is reminiscent of the modus operandi of phlebotomine sand flies that have used the method for at least 100 million years. Sand flies are one of the earliest groups of biting flies that developed a taste for vertebrate blood. They had probably evolved by the Jurassic, and early forms may have used their mandibles to obtain sap from primitive plants, much as some sand flies penetrate and obtain phloem sap from higher plants today.[143] Just when blood became the obligatory part of the diet is unknown, but by the Early Cretaceous, when the first fossils appeared in Lebanese amber,[174] some had possibly acquired this habit. They may have initially taken blood from wounds, but certainly by the mid-Cretaceous, the hematophagous habit was well established. It is difficult to say whether their bites would have caused raised itching papules or extended swellings on dinosaurs as they do in hypersensitive humans.[175]

While sand flies are known to engorge themselves on amphibians, reptiles, birds, and mammals, some prefer the blood of snakes and lizards.[176,177] Those that select reptiles have no problem inserting their mouthparts through the imbricate scales on their victims. Feeding on a snake takes about an hour, during which time an amount of blood equivalent to the insect's weight is ingested. This increased mass impedes flight until the serum portion of the blood is excreted. Then after taking three to five days to digest the meal, the female converts the remainder into eggs. Interestingly, males also crave blood even though they do not have the mouthparts to penetrate intact skin. Instead, they simply lap blood from wounds, including those made by biting females.[178]

Today, phlebotomine flies occur in a variety of climates from warm to tropical and habitats ranging from rain forests to semi-arid deserts.[175,179] Given Cretaceous temperatures,[180] sand flies would have been globally dispersed in a variety of environs. Adults might have preferred dark, humid sites on tree trunks and leaves or tree hollows, caves, and other excavations. As the climate in certain regions became drier, they adapted to micro-habitats such as rodent or turtle burrows that provided the humidity and temperature levels approaching those of their primeval homes. Phlebotomine larvae develop in concealed areas, like cracks and crevices in the soil, tree hollows and crotches, termite mounds, forest floor litter, and animal burrows.[175] Fungi may have been the basic food for the larvae, since two Cretaceous individuals were found associated with the fruiting bodies of a club mushroom[181] (color plates 8D, 12A). While small and fragile with a limited flight range, sand flies can travel up to a mile, and specimens found on islands suggest that wind currents carry the adults for much longer distances.[175]

Sand flies probably introduced trypanosomatids into dinosaurs (figs. 24, 35–37). This idea is based on the discovery of a specimen in Burmese amber that contains reptilian blood cells infected with the trypanosomatid *Paleoleishmania proterus*[254] (color plate 8C). Today, those phlebotomines in the genus *Sergentomyia* feed and transmit the related *Sauroleishmania* to snakes and several families of lizards.[183] The effect of trypanosomatid infections on reptiles has been little studied; however, when these protozoa were injected into four chameleons, they all died.[184]

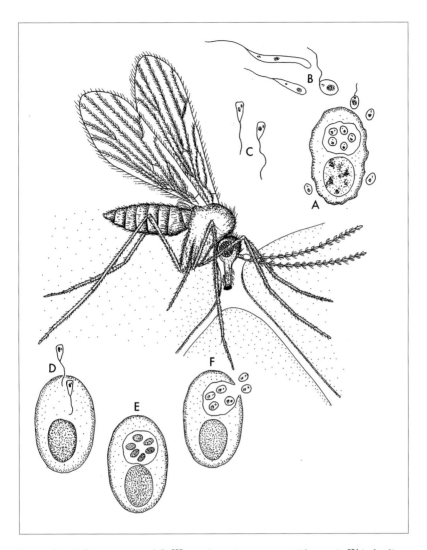

FIGURE 24. A Cretaceous sand fly[255] carrying a trypanosomatid parasite[254] is feeding on the exposed skin between the scales of a dinosaur. Upper right, stages of the parasite that occur in the alimentary tract of the sand fly: A. Infected dinosaur blood cell with small spherical amastigotes of *Paleoleishmania* in the gut of the sand fly. Isolated amastigotes can also occur in the blood meal. B. Amastigotes become elongate flagellated promastigotes that multiply by simple division (asexually). C. Short paramastigotes are formed that are infective to dinosaurs. Lower left, stages of the parasite that occur in the dinosaur: D. Two paramastigotes entering a dinosaur blood cell. E. Formation of a parasitophorous vacuole containing developing amastigotes. F. Dinosaur blood cell releasing mature amastigotes. Continue the cycle by going back to A. Dark oval objects in blood cells are nuclei. Not drawn to scale.

If dinosaurs were as susceptible to visceral leishmaniasis as humans are, entire populations probably were decimated. Human infections, which include both a skin and internal (visceral) form of the disease, are centered in Asia, Southern Europe, Africa and South America. Cutaneous leishmaniasis infects an estimated 12 million people throughout the world, and 1.5 to 2 million new cases appear each year.[153] The visceral form is called Kala Azar and was responsible for some 100,000 deaths in southern Sudan alone between 1989 and 1994.[185] American soldiers in Iraq and Afghanistan are familiar with the hazards of being targeted by infected sand flies. Probably close to a thousand have been infected with the cutaneous form of leishmaniasis, and a number now suffer from the potentially lethal visceral type. The standard treatment is pentavalent antimony—which has a high incidence of adverse reactions. Although medication represses the infection, flare-ups can occur throughout life.[159]

Fossil evidence indicates that vertebrate infections of *Paleoleishmania* were quite extensive. Of 21 fossil female flies examined, 10 contained trypanosomatids, an infection rate of nearly 50%. These figures indicate that *Paleoleishmania* had reached epidemic proportions in the Burmese amber forest 100 mya. Present-day levels of *Leishmania*-infected sand flies vary depending on geographical location and habitat. The causal agent of American visceral leishmaniasis occurred in only 0.05% of over 5,000 Colombian phlebotomines sampled and in just 0.23% of 860 Venezuelan specimens.[186] Cutaneous leishmaniasis is endemic in the Jordan Valley, and infection rates in sand flies collected from burrows of infected rodents varied from 9.2%[187] to as high as 56%,[188] which is roughly equivalent to that observed in Burmese amber.

Trypanosomes are not the only pathogens transmitted by sand flies. In the Cretaceous they also probably carried malarial pathogens to dinosaurs, since these insects are vectors of lizard malaria today, a highly infectious disease often resulting in death.[189] Sand flies also pass arboviruses to vertebrates. A human disease known as sand-fly fever is caused by a phlebovirus and produces flu-like symptoms.[175] Other arboviruses carried by phlebotomines include a reovirus (Changuinola virus) in sloths and a rhabdovirus in humans, wild carnivores, monkeys, and ungulates in South American.[175] There is evidence that Cretaceous sand flies were infected with polyhedrosis viruses,[172] which were feasibly the precursors of arboviruses carried by these bloodsuckers today.

Present-day sand flies also transmit bacteria and nematodes.[173] A disease called bartonellosis is caused by a bacterium that induces anemia (Oroya fever) or a chronic skin disease (Peruvian wart) in humans. If untreated, Oroya fever can kill up to 40% of its victims.[175] The potential effects of sand-fly transmitted viruses, bacteria, and nematodes on dinosaurs are difficult to ascertain, but certainly they could have been quite debilitating, especially under conditions of starvation or in individuals with a compromised immune system.

# 14.

## Mosquitoes

*The silvery rays of a bright moon silhouetted a small carnivorous dinosaur moving silently among the stalks of giant reeds bordering the forest. The night air was filled with the croaking of frogs, chirping of tree crickets, and rustling sounds of leaves in the wind. With large eyes adapted to low light intensities, the activities of a rodent-like mammal feasting on seeds in the undergrowth were easy to detect. The dinosaur crouched down and slowly began to stalk the unwary prey. When it heard the bending and snapping reeds, the quarry stopped chewing and immediately darted to the nearest tree. The predator followed in close pursuit and as the mammal began scurrying up the trunk, the dinosaur's grappling claws tore out some of the animal's hairs even as the prey managed to scamper high into the upper branches. The detached hairs drifted down into a blob of resin on the bark of the tree, leaving behind a lasting memento of this drama.*

*The commotion attracted the attention of squadrons of mosquitoes hovering in the vicinity of a nearby salt marsh. Most of the females had already laid one batch of eggs and now needed another blood meal for a subsequent one. As the dinosaur peered hopefully upward at what would have been a meal, a cloud of sanguinary mosquitoes alighted unobtrusively on its body and began probing the thin skin separating the tuberculate scales. Most settled on the back and shoulders, while a few selected areas around the eyes and ear openings. With surgeon-like precision they began piercing the capillaries and sucking up blood. The bites were ignored and after three or four minutes, the insects withdrew their stylets and laboriously flew to some lower branches to digest the meals. The on-*

*slaught was to continue throughout the night hours as others avidly took their places.*

Mosquitoes are uncommon as fossils, even in recent amber from the Dominican Republic,[190] and only a single uncontested specimen has been described from the entire Cretaceous[191] (color plate 11D). Another possible representative from Early Cretaceous deposits of England[35] suggests that these bloodsuckers occurred throughout the period.

By the Late Cretaceous, their hosts would have included mammals, birds, reptiles, and amphibians, as well as dinosaurs. Primitive species could have developed in salt or brackish water and preyed on vertebrates in the same habitat, similar to those that now attack sea turtles in Florida and South America.[192,193] The bloodsuckers seemingly arrive out of nowhere when the first turtles appear, then swoop down and eagerly feed on the hapless giants as they laboriously dig holes and deposit their eggs. The stoic turtles show no signs of irritation from the hundreds of ravenous insects simultaneously imbibing their blood. When satiated, the mosquitoes retreat to various dune plants and rest until their meal is digested. This same scenario probably was played out innumerable times along the ancient coastlines of epicontinental seas, including those near the Canadian amber site.

Just being attacked by massive numbers of them can be dangerous. When enormous swarms descend on humans, half of their blood can be removed in just two hours.[194] A small dinosaur hatchling could have been exsanguinated in less time. At least five genera of mosquitoes attack lizards today. Most of these are opportunists, meaning that they also dine on man and other mammals, birds, and amphibians.[195–197] But more important than removing bodily fluids are the microorganisms they add to the circulatory system of their victims. These hematophagous insects appear to have no problem penetrating the scaly surface of reptiles, and no trouble leaving and acquiring various microorganisms.

Malaria is the single most important disease carried by mosquitoes, not just to humans but to birds and reptiles as well. Some 1.6 billion people are at risk and about a million die each year from the disease.[194] Could malaria have played a role in the disappearance of the dinosaurs? Mosquitoes currently transmit some 29 species of *Plasmodium* malaria to reptiles, but the infections appear to be tolerated.[197-200] However, in the Cretaceous, when arthropod-borne malaria was a relatively new disease, the effects on dinosaurs could have been devastating until some degree of immunity was acquired.

I can personally vouch for the ability of mosquitoes to both torment and infect victims during the dead of night. When out searching for medically important insects in the Ivory Coast, West Africa, our research team spent many nights in the bush. We each carried a hammock equipped with mosquito netting on all sides. If, after securing both ends to tree trunks, you could enter quickly enough to leave the mosquitoes behind and you didn't mind the constant humming of the hungry females right next to your head all night, then perhaps you might get a few hours sleep. But if the oppressive heat and humidity made you toss and turn, and your sweaty arm touched the mosquito netting, you awoke the next morning with twenty to thirty bites, evidence of a mosquito banquet. That is exactly how I became infected with malaria and learned the drug I was taking (chloroquine) was not very effective against newly mutated strains the mosquitoes were carrying. Enervating bouts of fever and shivering showed me how serious this disease was, and without further treatment I may have become one of the million or so fatalities this disease claims every year. With such a high mortality rate, it was understandable how the sickle-cell trait, which conveys some immunity against malaria, became incorporated into the African population. Just how many of our laboratory workers had this gene was impossible to know, but they certainly were able to survive their malaria attacks, which at most caused them to remain at home one or two days every fortnight.

Scores of arboviruses are presently carried by these bloodsucking pests,[194] and some of them have a great capacity for mutating and forming new virulent strains. West Nile virus was introduced into the New York area in 1999, and by 2004 had spread across North American, killing incalculable numbers of birds, mammals, and reptiles along the way.[201,202] Birds, who appear to be especially susceptible to the virus, are considered to be the major dispersal agents. Symptoms in humans include high fever, internal bleeding, encephalitis, and death. To make matters worse, once established, infections can be spread by animals eating virus-contaminated meat. In one case, over a thousand captive alligators died after eating meat tainted with West Nile virus.[202] Considering the potential mutability of the West Nile virus and both the insect and oral transmission routes, we can expect a still-wider range of vertebrates to be infected. Especially when so many opportunistic mosquitoes that feed on a smorgasbord of reptiles, amphibians, birds, and mammals could help spread the disease. Fortunately, the West Nile virus is not presently highly virulent for humans, since if it was, the number of deaths could equal those of the plague during the Middle Ages.

One of the most notorious arthropod-transmitted viruses causes yellow fever. This pathogen is so lethal that if we were unable to control the vectors or vaccinate the population, there would be large uninhabitable areas throughout the globe. Yellow fever virus is restricted to primates and apparently was acquired from African monkeys, which incidentally do not show symptoms.[194] When untreated, mortality reaches 80% in some areas in Africa, and even with a vaccine, it is estimated that up to 200,000 new cases occur each year. Similar types of arboviruses undoubtedly wreaked havoc with dinosaurs, causing epidemics that debilitated entire populations.

Cretaceous mosquitoes also theoretically transported parasitic nematodes, especially those causing diseases known as filariasis. As with similar nematodes in lizards and crocodiles today, the parasites probably entered the dinosaur's connective tissues, muscles, lungs, heart, blood vessels, or skin. An impressive number of filarial nematodes carried by mosquitoes now infect birds and mammals.[173] One that lives in humans collects in the lymphatic system and can cause enormous swelling of the limbs and other body parts, resulting in a disease called elephantiasis.[194] The best-known filarial nematode in the temperate zone is the dog heartworm, which lives in the pulmonary artery and heart. The wide range of animals that are parasitized by mosquito-borne nematodes suggests that in the past dinosaurs were similarly infected. Although it is difficult to say just what symptoms they would have had, filarial worms can seriously damage the viscera of reptiles.[203] If dinosaurs suffered as much as modern mammals do, the resulting illnesses were quite incapacitating, leaving the victims easy prey for carnivores.

Presently, mosquitoes occur throughout the entire world (except Antarctica) and survive under a broad range of climatic regimes and habitats. While the great majority require a source of standing water for the development of their larvae, these habitats can range from large lakes, ponds, and small enclosed pools in tree holes and rock formations to miniscule amounts of water in the bases of pitcher plants, epiphytic bromeliads, and even leaves. It seems that if there is any water around, there is a mosquito that can use it as a breeding source.[194]

Most engorge at night or around dawn and dusk; only a few are diurnal. They locate their prey by following carbon dioxide emissions or the release of oactic acid and octenol. Visual detection is particularly important when foraging in daylight, especially for those that attack in open areas. Some species feed in many different habitats, while others select wooded or unobstructed terrain or remain close to breeding sites. A few seek specific hosts while certain groups are opportunistic and attack victims belonging to three or more vertebrate classes. Within the amber forests, mosquitoes subsisted in all strata. There would have been no relief for dinosaurs since under the tropical and subtropical conditions of the Cretaceous, these bloodsuckers in all likelihood were active throughout the year.

# 15.

## Blackflies

*A mountain stream coursed around and over boulders and tree roots through a glen in the forest. Clinging to the surface of the rocks were clusters of small, upright blackfly larvae equipped with a pair of strainers used to sieve out passing food items from the current. Unlike the robust adults that aggressively searched for blood, these immatures quietly waited for miniscule food items to pass by and be netted. When the blackfly larvae were satiated and eager to make the transformation into the air, they formed pupae attached to the rocks in the stream. A few days later, an adult blackfly emerged from each pupa. They didn't mind experiencing a few splashes of water before they became airborne. Masses of these tiny ferocious insects emerged at the same time and formed a swarm that flew off in search of their first blood meal.*

*A group of hadrosaurs browsed among the reeds bordering a pond. One of the more inquisitive members detected floating white flowers on a mat of water lilies and ventured into the water to dine on the treat. Thick, sticky mud oozed between large toes as the big female shifted ponderous weight from leg to leg. Upon reaching the tasty plants she bent forward to the water's surface and began stripping off the flowers and lush, succulent leaves of this delicacy until only the tasty bulbs half buried in the mud were left. After some wary glances around the edges of the pond, the duckbill submerged a massive head and began prodding the bottom with broad sensitive lips. Mayfly and water-beetle larvae frantically scurried away from the disturbance while caddis flies pulled into their cases and remained motionless. At this moment, a famished horde of blackflies landed on the back of the dinosaur and started to penetrate the thin layer of skin with their sharp mandibles. The unfortunate beast reacted to the vicious stabs by jerking her head out of the water and attempting to brush the pests away. Any vertebrate could tolerate some suffering from these insects, but not that generated when so many attacked at once. By now the voracious insects were excising miniscule portions of flesh and lapping up the exposed blood. Snorting wildly, the frantic hadrosaur lay down in the water, but the pain persisted and the blackflies remained attached. Rising and shaking vigorously, the harried animal splashed out of the pond and fled toward the forest, stumbling over some boulders and bruising both knees as she attempted to dislodge the tormentors. Risking a broken leg in this hasty flight, she finally found refuge among some tall ferns. The blackflies, if undisturbed, would have continued feeding, and if their numbers were thick, could have drained her body of blood.*

Did millions upon millions of blackflies soar over Cretaceous fens and fern grottos, clogging the nasal passages of feeding dinosaurs? Would their appearance cause dinosaurs to stampede, tripping over rocks, breaking legs, or plunging off cliffs in a frantic rush to escape the torturers? Were toxic substances in insect saliva causing allergic reactions and blood poisoning? Could masses of these noxious flies cover a dinosaur's skin and remove so much blood that it would expire from exsanguination? All of the above are known to occur today when blackflies attack grazing bovines on the plains of western Canada.[204,205] These insects are capable of swarming over the prairie for miles searching for cattle. The most dramatic result of these forays is exsanguination—when hundreds at a time settle on a victim and drain the blood over the course of a few days. Even if some survive the initial attack, many eventually succumb from broken bones, starvation, and infections. Such a scene is not limited to Canada; a few decades ago, blackflies undertook massive migrations along the Danube in eastern Europe that resulted in the death of thousands of cattle, sheep, goats, and pigs.[205] This may well have happened to dinosaurs.

Blackflies were certainly present in the Cretaceous since records of them extend back to the Jurassic. Fossil remains occur in Europe, Australia, Asia, and North America.[205–208] It would be helpful to establish what Cretaceous species dined on since there is no evidence that this group is attracted to cold-blooded animals today. A single study reports females clustering around the eyes of a turtle, but not feeding.[209] In fact, blackflies are the only large group of hematophagous insects that have not been observed feeding on reptiles.

However, they do attack birds and mammals. While there were Cretaceous birds and mammals, it seems unlikely that these bloodsuckers would have ignored the dinosaurs because they were an obvious unlimited source of protein. Furthermore, if dinosaurs and birds are as closely related as some believe, then dinosaurs might well have been among their original hosts.

It has been suggested that their sanguinary pursuits started with pterosaurs, a group that may have been warm blooded because they needed heat for their flight muscles.[205] Did pterosaurs, some of which were sparrow sized, pre-adapt blackflies to eventually partake the life's fluid of birds and mammals? Then again, whether dinosaurs were warm- or cold-blooded or some condition in between may not have mattered to blackflies. If some lineages specialized just on dinosaurs, they probably disappeared along with their hosts, leaving behind those that favored feeding on birds and mammals.

One of the blackflies preying on hadrosaurs or their kin could have been carrying a malarial protozoan (*Leucocytozoon*) in the salivary glands, which was then introduced into the bloodstream of the victim. In the following weeks, these microorganisms would multiply in the tissues and blood cells of the host, establishing a chronic infection that could weaken the animal's immune system, lower its reproductive potential, and even cause death. Today, a similar disease (leucocytozoonosis) is often fatal to ducks, geese, chickens, and especially turkeys.[204]

Cretaceous blackflies also may have been transporting filarial nematodes. These wiggling parasites would enter the blood vessels and be transported into the internal organs. After accumulating in certain tissues, the nematodes would mate, deposit their eggs, and the young microfilaria would then wait to be picked up and transported to a new host animal by yet another blackfly. This cycle is much like one that currently infects ducks.[173]

A particularly grim blackfly-transmitted nematode similar to the one that causes river blindness in humans feasibly may have been infective to dinosaurs. Some 17 million people in tropical Africa and South America are afflicted with this blinding disease, which is caused when larval nematodes spread into the eyes of the victim, resulting in visual impairment and eventually total loss of sight.[204] A dinosaur with compromised vision not only had difficulty finding food, but also was easy prey for predators.

While studying the nematodes causing river blindness in West Africa, our team of scientists found that an effective way to trap black fly vectors was to wait along the side of a stream and catch them as they settled on our skin. We were highly motivated to act quickly, since if they bit us, within seconds the microscopic nematodes the flies were carrying would enter our bloodstream—and no one wanted to risk losing their sight. Apparently we were not fast enough, and one of the group eventually developed telltale nodules that indicated the presence of mating adult worms, whose progeny is responsible for vision loss. Treatment at that time consisted of two drugs, one for the juvenile nematodes migrating through the tissues and another for the adults in the skin lumps. Since both of these remedies had side effects, our colleague had the cysts surgically removed in Europe. Recently an ivermectin-based drug has provided a much more effective and safer type of treatment, and hopefully will aid in the extermination of this tragic disease.

With the exception of Antarctica, blackflies are presently distributed throughout the globe. Their larvae require running water for survival, so they are limited to locations with streams and rivers. However, the adults are capable of flying hundreds of miles (assisted by the wind) from their breeding sites in search of blood, and thus their range would have been fairly extensive in the Cretaceous.

# 16.

## Horseflies and Deerflies

*A rather large horsefly rested on a ginkgo leaf while the early rays of the sun dried its new brown-spotted wing membranes. Just a few days before, this insect had been a white legless grub crawling through mire on the shore of an extensive lagoon, sucking the lifeblood from any luckless invertebrate it could overpower. Now a second phase in life was beginning. The large, iridescent eyes searching the landscape for prey gave the illusion of beauty and innocence but masked a rather sinister passion—a lust for vertebrate blood. Multiple lenses in the eyes suddenly registered the movements of an iguanodon feeding on ferns along the edge of the marsh. The horsefly immediately became airborne and flew directly toward the dinosaur. Initially delaying contact, the ravenous fly circled around the animal's head while looking for a safe landing spot. The flight pattern shifted into smaller and smaller circles until, almost effortlessly, the insect settled on the large vertebrate's shoulder. Exposing the powerful mouthparts, the fly began to gash an opening through the victim's skin. The searing pain caused the herbivore to momentarily stop feeding and turn her head. But she couldn't reach the irritation and twitching the skin failed to dislodge the fly. No amount of the jerking could disturb this determined predator, and soon a sanguineous fluid began to trickle down between the tuberculate scales as the insect's abdomen took on a crimson hue. The smell of fresh blood attracted non-biting flies that welcomed this free meal of concentrated protein.*

*Then, almost as silently as the landing of the horsefly, the first sign of an even more sinister danger approached from the water, disguised in a series of bubbles and small whirlpools surrounding two protuberant eyes and a set of nostrils. Without a second horsefly bite on the hind leg, the agitated dinosaur might not have thrown caution to the wind and splashed into the brackish waters to dislodge the intruders. While the soaking finally dislodged the fly on her leg, the plunge took her further out than she would have normally ventured. The attack by the lurking giant crocodile was swift and deadly. Vise-like teeth clamped down on the young iguanodon's haunch and the helpless animal was dragged into deeper waters. If horseflies hadn't badgered the animal, thus changing her normal course of cautious behavior, the iguanodon's life might have played out differently.*

Horseflies and their smaller cousins deerflies of the family Tabanidae were widespread throughout the Cretaceous and because they now feed on both warm- and cold-blooded animals,[210–213] they certainly took blood from dinosaurs (fig. 22). Their painful bites and persistence behavior would have greatly irritated the huge reptiles, just as they do many animals today, from reptiles to humans. At least four species of horseflies are known to prey on crocodiles and anacondas in the Amazon.[214] The onslaught begins with the flies circling rapidly around the victim before alighting. Feasting commences as soon as they land, and each species appears to have a preferred feeding site, either the head or back.

The wounds are obviously unbearable since caimans become extremely agitated and wave their tails in a vain effort to rid themselves of the pests. Because that action is usually unsuccessful, the reptiles resort to diving into the water, a recourse also used by anacondas who are attacked. While this gives some temporary relief, the insects just wait around for the reptiles to emerge and then resume the assault. Even when the victims are partially submerged, the bloodsuckers will attack any exposed body parts. One wonders if many of the large sauropods that are depicted as feeding in water may have adopted this behavior to keep biting flies away from their ventral surfaces.

Turtles, including the large Galapagos Island tortoises, are also victimized by horseflies.[215] The flies seem to know when leatherbacks are going to come out of the sea to lay their eggs. At least six species of tabanids feed on marine turtles in Suriname and French Guiana, attacking the new arrivals on necks, shoulders, and legs from sunrise to sundown.[216] Freshwater species are likewise ravaged during the summer months, and to add injury to the insult, these flies also transmit turtle malaria.[217] Perhaps the most vulnerable animals are mammals. Not only man but also his domestic animals can be savagely bitten by these persistent creatures. Just moderate to heavy feeding can result in weight loss in beef cattle, reduced milk production in dairy cows, and hide damage (for leather products) from the resulting large punctures.

Tabanids infect victims with pathogens by a simple, direct method called mechanical transmission. Microbes picked up on the flies' mouthparts after feeding on a diseased animal are subsequently transmitted to a healthy animal via the wounds. The mouthparts function like a pair of scissors that were dipped into a solution filled with germs. Using this simple method, they are able to transmit viruses, rickettsia, bacteria, and protozoans to a range of animals. Among the pathogens transmitted are anthrax bacteria, spirochetes such as those responsible for Lyme disease, and viruses causing leukemia and cholera.[218] This method of transmission is probably how many pathogens were originally spread to vertebrates by insects.

If a horsefly's mouthparts were contaminated with bacteria from the last meal and some of the cells entered a wound, they might have started to multiply in a dinosaur's bloodstream. Whether the infection would continue to spread, eventually causing a lethal septicemia, would depend on the nature of the bacteria and the state of health of the animal and its immune system.

The only parasites carried by tabanids that actually develop inside the fly are filarial nematodes. Loiasis, a potentially debilitating disease of humans and simians in Africa, is just one of these. The adult worms live under the skin near the head of the victim, sometimes passing under the conjunctiva of the eye, which is how they became known as eyeworms. Another filarial nematode lives in the heart and blood vessels of elk and moose, clogging their arteries and causing blindness and even death.[173,218]

Thus bloodsuckers could have left tiny parasitic nematodes in dinosaur wounds. In time, the roundworms would develop, reproduce, and bear their young in the tissues of iguanodons and others. They probably would not kill their host, but they would become yet another physical burden the dinosaur would have to endure during its lifetime.

So a young iguanodon might not have just borne the feeding scars of horseflies, but could have been infected with any number of pathogens they introduced. Most Cretaceous dinosaurs would have been fair game, especially the large, slow-moving sauropods that had no real protection against biting flies. Since tabanids are strong fliers, they probably occupied the same wide range of habitats during the Cretaceous they do now.[218]

# 17.

## Fleas and Lice

*A warm rain had caused tens of thousands of termites to swarm out of a nest at the base of a rotting log, and a feathered juvenile theropod, a coelurosaur was frantically rushing after them, savoring their rich, fatty bodies. After leaping into the air to snap at those that had just become airborne, the youngster suddenly stopped and began frantically scratching among the feathery structures on his neck. Residing there were two species of strange, wingless insects. One, a large, partially flattened louse almost a half-inch in length, remained fairly quiescent between the feather bases close to the skin surface. From time to time, the creature's vertical mandibles scraped away and devoured some of the epidermis, which served as its major food source. This giant louse was one of many that lived here, and the resulting irritation from their bites had caused the dinosaur to stop and claw at the itchy infested area. But the disturbance only made the primitive lice extend flattened antennae and secure themselves more firmly to the host.*

*A second, somewhat smaller parasite living in the downy plumage of the small coelurosaur had inserted a short, sturdy beak into the animal's skin and was sucking up blood. This primitive flea had stretched out monstrous hind legs and used them to grasp the feather bases for stability. Earlier, these same muscular legs had been employed to propel the flea onto the victim's body. A series of other appendages protruding from the abdomen provided additional attachment. An entire community of fleas and lice infested the youngster, having crawled aboard when he hatched, and succeeding generations would continue to reside there until his death.*

Parents cringe when their child brings a note from school informing them that a lice infection has been detected in the classroom and their child needs to be treated. And pet lovers know all about the astonishing jumping abilities of fleas and their painful, itching bites. Such pests have followed us into urban communities, and even with all our efforts to control them, still manage to plague us. Think how primitive man must have suffered!

Where would these pests have occurred on dinosaurs? A reasonable location would be among the plumage of feathered dinosaurs. Fleas and lice thrive under the protection of hairs and feathers on both mammals and birds. Plumed dinosaurs such as *Sinosauropteryx*, a small coelurosaur, were probably ideal hosts for early fleas and lice.

While modern fleas can be enticed to take blood from lizards,[219] they do not seek out reptiles in nature, probably because there is little place for concealment. However, flea-like creatures have been described from the Mesozoic that could have attacked ancient reptiles. One of these is the outlandish *Strashila*, equipped with crab-like claspers and long claws.[104] This small creature had extremely powerful hind legs that dwarfed the rest of the body (fig. 21). We sometimes are able to surmise the habits of extinct organisms by their structural characters, a principle known as functional morphology. A robust beak indicated that *Strashila* may have sucked blood, and clasping processes on the hind legs might have been used to grip feathers. It was suggested that these creatures sucked blood from pterosaurs, but they could just as well have attacked feathered dinosaurs. This bizarre lineage probably became extinct sometime in the Cretaceous.

Another bewildering Mesozoic flea-like insect is the large-bodied *Saurophthirus*. Their prolonged proboscis, extended claws, and long legs have been considered modifications for feeding on pterosaur wings.[221,222] The peculiar legs also could have been

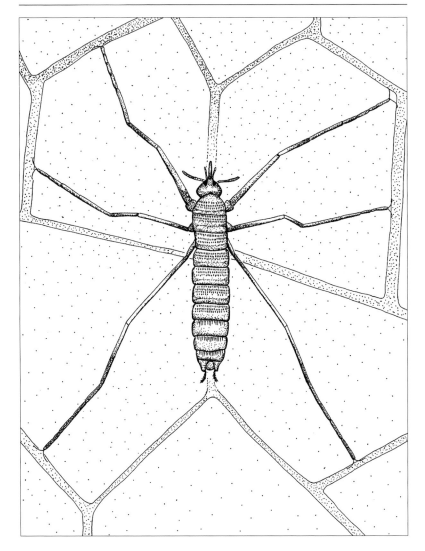

FIGURE 25. The long-legged *Saurophthirus longipes*[221] is ready to take a blood meal after securing its legs to the edges of plate-like scales on a dinosaur.

used to grasp the edges of dinosaur scales while feeding on blood with the extended proboscis (fig. 25). Long-legged fleas may have been widespread in the Cretaceous, since sedimentary fossils have been discovered in Australia.[206]

A contingent of these ancient fleas may have attacked dinosaur hatchlings, retreating in the debris under the nest to lay eggs. If Cretaceous dinosaurs had permanent or even semipermanent resting areas, these made ideal domains for flea larvae. Extant fleas lay their eggs in the sleeping and resting sites of their victims, and the larvae develop on hair, feather, and skin remains. Then again, it is possible that a few flea larvae actually developed under the skin of dinosaurs, just as some do today on mammals.[220]

Ancestors of the sticktight flea could have pestered dinosaurs. These insects attach themselves to the skin around the head of various birds, causing anemia when populations are high. Scratching at those affixed near the eyes can result in ulcerations and even blindness in chickens.[220]

Modern fleas are capable of transmitting a number of pathogens to warm-blooded animals, the most notorious being the bacterial organism that causes plague. Humans are infected from bites of fleas that are carrying the plague organism or by handling infected animals. Once established in the body, the infection can invade the internal tissues, including the lungs. When victims cough or sneeze out the pathogens, others can become infected, thus establishing a pulmonary type of the disease. Plague epidemics, which have occurred at three periods in recorded history, have killed more humans than all wars combined. The first pandemic broke out in Africa in the sixth century A.D., and after spreading though southern Europe was responsible for an estimated 40 million deaths. The second outbreak, known as the Black Death, originated in Asia in the fourteenth century then moved westward over the next two hundred years, eventually killing some 25 million Europeans. A third pandemic originated in China's Yunnan province (near the Burmese amber mines) in the mid-nineteenth century and continued to the mid-twentieth century, being carried by infected ship rats around the globe. This pandemic was responsible for approximately 20 million deaths in India alone.[220] Did dinosaurs suffer from plague or plague-like epidemics spread throughout their populations by ancient fleas? Could such microbes have circulated among them via flea bites or inhalation, as occurs in plague outbreaks now?

Cretaceous fleas possibly mechanically transmitted viruses to dinosaurs just as present-day fleas do the myxomatosis virus of rabbits. This virus-flea combination was recently used to reduce the flourishing populations of European rabbits that had been introduced into Australia.[220] The virus enters the tissues and causes spreading skin lesions that eventually kill the victims. A virus-flea combination would have been an excellent way for viruses to disseminate among dinosaur hatchlings living in crowded conditions or populations traveling in herds.

Lice are highly specific parasites of mammals and birds, and are placed in two categories, the biting (Mallophaga) and sucking (Anoplura) types. The latter all feed on mammalian blood currently while nourishment for the biting forms includes feathers, hair, and skin, as well as blood. The more primitive types that include birds among their hosts today are the biting forms, and this is the kind that bothered our feathered dinosaur. The habit of biting and scraping at the base of the plumes would have caused skin irritation and bleeding as well as feather loss, similar to louse infections in present-day birds. If the parasite populations continued to build up, there would have been the possibility of dermatitis and anemia, not to mention infections from the open wounds.[223]

Fossil lice are rare; however, the giant *Saurodectes* from the Early Cretaceous of Transbaikalia could have fed on pterosaurs or feathered dinosaurs,[224] and if warm blooded, as some believe, then they most certainly would have been susceptible to lice infestations. Even if dinosaurs were cold blooded, the body temperature of cold-blooded animals remains fairly high under tropical conditions. These archaic lice might have evolved on feathered dinosaurs before moving on to birds and mammals.

Even though bird lice are not known to transmit many pathogens, some can act as reservoir hosts. Individuals that feed on birds dying from a rickettsia disease known as fowl cholera retain the pathogens in their digestive system for six days, long enough to transfer them to healthy animals.[225] On the other hand, mammalian lice do vector several important human pathogens, and if sucking lice existed in the Cretaceous, they might have transmitted similar bacteria and rickettsia to dinosaurs. Whether any of the possible Cretaceous louse-borne pathogens were as serious as louse-borne human typhus is difficult to say. This rickettsial disease, carried by a sucking louse, was introduced into North America by the Spanish in the sixteenth century and killed some 2 million Native Americans. The pathogen, which is transmitted when infected lice feces are scratched into the skin, has claimed many lives throughout history, with outbreaks usually killing from 10% to 50% of the victims.[223] This shows how a disease can be spread just mechanically by a vector, since the victims themselves scratch the louse-borne microbes into their skin.

Some lice carry the intermediate stages of filarial nematodes. Among these are species of roundworms that lodge in the neck region, heartworm parasites that live in the myocardium, and a nematode that occurs in the tendons and muscles of the foot joints of birds.[173] When the latter die, the birds may develop arthritis of the joints, affecting leg movement. If dinosaurs were infected with similar nematodes, carnivorous ones might have had difficulty catching their dinner and herbivorous ones would have had problems evading predators.

The associations we have proposed between dinosaurs, fleas, and lice are highly conjectural. There were certainly adequate numbers of birds and mammals in the Cretaceous forests to sustain these insects, but it is also reasonable to assume that at the very least, feathered dinosaurs could have been early hosts to primitive fleas and lice.

# 18.

## Ticks and Mites

*A miniscule tick larva sat motionless on the arched frond of a large tree fern. The young juvenile had recently emerged from a cluster of eggs laid in leaf litter on the forest floor. Driven by the basic instinct to obtain a blood meal, the hatchlings had scattered in various directions. They all eventually ascended an assortment of plants to reach a height suitable for contacting a passing vertebrate. Some had chosen horsetail stems, while others had selected tree trunks. The journey for a few ended when they became entrapped in beads of kauri resin. The tick on the fern leaf waited patiently, since each larva could survive for weeks without food, and aimless wandering would only use up energy. Luckily a group of pachycephalosaurs soon began browsing on the fern foliage. Only glimpses of their bodies appeared among the shaking plants, although now and then a head would suddenly emerge when one reached upward to tear open a trunk and access the tasty fiber within. The vigilant tick was alerted by all the commotion, and sensing the presence of potential hosts, raised short front legs in anticipation and waited for that critical moment when one of the quadrupeds brushed against the frond.*

*When that occurred, there could be no hesitation, and the larva quickly dropped onto the thick, pebbly skin. For a first meal, the tiny arachnid chose an area of soft tissue around the ear opening. Using deft palpal claws as attachment devices, the tick inserted a barbed tube into the dinosaur's skin. Viscous secretions in the saliva cemented the tick's mouthparts in place and the blood meal began. When satiated, the tick waited until dusk, then dropped to the ground and crawled under some dried leaves at the base of the tree fern. Soon the larva would molt into a nymphal stage, scale another plant, and await another passing dinosaur. Molting after the second meal into an adult female meant the subsequent feeding would be a longer one, furnishing protein needed for egg produc-tion. The large volume of blood consumed at that meal would swell her body up to ten times the original size. During a lifetime, she would deposit several thousand eggs, ensuring survival of the species.*

While ticks and mites are not insects but arachnids related to spi-ders, both probably played important roles in the transmission of pathogens in the Cretaceous world. There are hard and soft varieties of ticks, and in spite of the difference in body shape, texture, and feeding habits, both sexes of all species require ver-tebrate blood for survival. After inserting a hollow tube, or hy-postome, into the integument, blood is pumped directly into their stomachs. Ticks are very patient and will remain on vegeta-tion for days, weeks, or even months until a host passes close enough for attachment and feeding. A few have elected to live in or near the domiciles of potential victims, where a blood meal is readily available.

Mid-Cretaceous records of both hard and soft ticks[45,226] show these arachnids were a vital part of that ancient world. While the Cretaceous soft tick appears to have utilized birds, the hard tick had features of extant reptile-feeders and probably consumed di-nosaur blood (color plate 11E). The claws on its palps, a unique feature unknown in today's ticks, presumably were used for grasping tuberculate scales.

Presently, ticks parasitize birds, mammals, reptiles, and am-phibians. Dining on reptiles is widespread, and seven genera have representatives attacking tortoises, snakes, and lizards.[227] Some of the most bizarre associations involve *Amblyomma* ticks that engorge on pelagic sea snakes and marine iguanas.[228] Primi-tive *Aponomma* ticks live almost exclusively on reptiles, and one species positions itself on the forelegs, between the toes, in the olfactory sacs, around the cloacal opening, and in the groin of monitor lizards. Another aponommid feasts between the scales, in the eye sockets, and on the head and neck of pythons.[229]

The saliva of some ticks is poisonous and can cause inflammation, a fatal paralysis, or even anaphylactic shock.[230] These neurotoxins are released from the salivary glands and enter the blood of the victim while the tick is feeding. In humans, the paralysis starts in the lower extremities and moves upward to the head region. Walking becomes difficult and the victim becomes easily fatigued as the first signs of paralysis occur. The numbing effect continues and affects the tongue and facial muscles, eventually causing total body paralysis. Without removing the tick, convulsions and death from the shutdown of diaphragm muscles controlling respiration may follow. Similar reactions in dinosaurs, especially among the very young, would have left the animals completely helpless and easy prey.

While feeding on pachycephalosaurs or other dinosaurs, a tick larva may have ingested some bacterial spirochetes that were circulating in their blood. If so, these microorganisms could have multiplied and accumulated in the tick's salivary glands. When taking a second meal, some of these bacteria would be transferred into the new victim and carried by the blood to various body organs. For a brief time, there would be no effect from the multiplying bacteria. However, the animal soon would become sluggish, with stiffening legs and lethargic movements and unable to keep up with the rest of the group during their daily treks to the fern grottos, eventually falling victim to a flesh eater.

Ticks have a bad rap, and rightfully so since they carry some very dangerous pathogens. With an extensive host range, these arachnids are considered the second-most important vectors of human disease after mosquitoes,[231,232] and have been transmitting pathogens to humans and domesticated animals for thousands of years. Ticks became widely known to the public when they were recognized as vectors of Lyme disease, induced by a spirochete acquired from feeding on diseased rodents or deer in North America and sick birds and rabbits in Europe. Because of severely debilitating effects on humans and the sudden occurrence throughout the United States, Lyme disease has been intensively studied. The first symptoms are often a circular expanding rash beginning around the tick bite. This is accompanied by fever and joint and muscle pains, similar to those felt during the onset of flu. Once the bacteria have become established, nerve inflammation in the joints can lead to difficulty in walking. Diseased animals or those carrying contaminated ticks can spread the infection over considerable distances.[230] It is easy to visualize such tick-borne diseases conceivably spreading through dinosaur populations in a similar manner. The intense fatigue associated with these diseases would put infected animals at risk.

Reptiles and birds have their own varieties of Lyme disease carried by ticks, but some lizards are protected from the infection by a substance in their blood that destroys the spirochetes.[233] Such immune responses could have evolved in Cretaceous reptiles challenged by infections of ancient spirochetes.

Other bacterial pathogens transmitted by these arachnids cause illnesses known as ehrlichiosis in livestock and dogs, and tularemia in mammals, birds, and amphibians.[230] Ehrlichiosis is often fatal to dogs, especially susceptible breeds like German shepherds. Additional tick-borne pathogens include protozoans that cause malarial-like ailments in birds and mammals. These parasites develop in circulating red blood cells and cause fever, sweating, headaches, and chills, and infections often have fatal results. Cattle acquire the disease from ticks that gorged on infected deer (reservoir hosts), and while deer do not appear to suffer, cows can enter lethal comas. Other malarial-like organisms vectored by ticks are responsible for African East Coast fever that is lethal to cattle and buffalo.[230]

Ticks also spread a number of rickettsia, such as Rocky Mountain disease that causes death in humans.[230] Still another tick-borne rickettsia that survives in soil and water for months causes Q fever in birds, mammals, and reptiles, including lizards, snakes, and tortoises.[234,235] People acquire the infection by handling diseased animals or items contaminated with tick feces and may suffer fever, chills, and even fatalities.[230]

Viruses carried by ticks cause encephalitis and hemorrhagic fevers, leading to death in a wide range of mammals, including man. Reptiles may play a major role in the disease cycle by being significant reservoir hosts for tick-borne viruses.[230] And finally, some African and South American ticks spread filarial parasites to mammals.[173]

The number of deadly and debilitating pathogens that ticks transmit is astonishing, and there is no reason why dinosaurs would not have been targeted. These aggressive biters are at home on warm- or cold-blooded animals, and no amount of scales, feathers, or hair can impede them. They undoubtedly were quite annoying when large numbers clustered around ear openings, settled under the eyelids, or entered the nostrils. They occur throughout the world in a variety of climates and habitats from tropical forests to deserts, which include conditions found throughout the Cretaceous.

On the basis of our present knowledge, perhaps just as many mites preyed on dinosaurs, since many of the types found today on vertebrates and invertebrates lived in the Cretaceous (color plates 11F, 13B, 14C). In various parts of the world, mites cause a condition known as mange in mammals and birds. Mange mites are adapted to living on the legs of birds in spite of the scaly coverings, and "scaly-leg mites" even cause lesions on the keratinized leg portions of game birds.[236] Could the ancestors of these have victimized dinosaurs too?

Of those that attack reptiles, one type exploits geckos, alligators, and rock lizards, while another type prefers skinks and terrestrial lizards.[237] Certainly, early members of these mite families would have accepted dinosaurs as hosts.

All stages of vertebrate-parasitic pterygosomid mites take lizard blood,[238] and they remain on the host during their complete developmental cycle from a 6-legged larva to an 8-legged adult. Only the females leave the host to deposit their eggs in the surrounding area, and upon hatching, the tiny active larvae secure themselves under scales on the legs, neck, or back.[239] Female mites of this group attacking iguanas in Brazil are flattened so they fit snugly beneath the lizard's scales.[240] Even after the reptile dies, the parasites will remain attached for weeks, feasting on the carcass until it is completely mummified.

Chiggers (trombiculids) are parasitic only as larvae, then become predators on small invertebrates as nymphs and adults.[241] These mites transmit malarial parasites to rodents and lizards,[236] as well as filarial nematodes to mammals.[173] They certainly existed during the Cretaceous, and the dinosaurs couldn't have escaped their attacks.

Another group that existed back then was the beetle mites, or oribatids. Widespread today on plants, they serve as intermediate hosts of tapeworms, which sometimes cause convulsions and death in humans with heavy infections. It is possible that herbivorous dinosaurs were infected with mite-borne tapeworms, but whether they experienced such severe reactions will remain unknown.

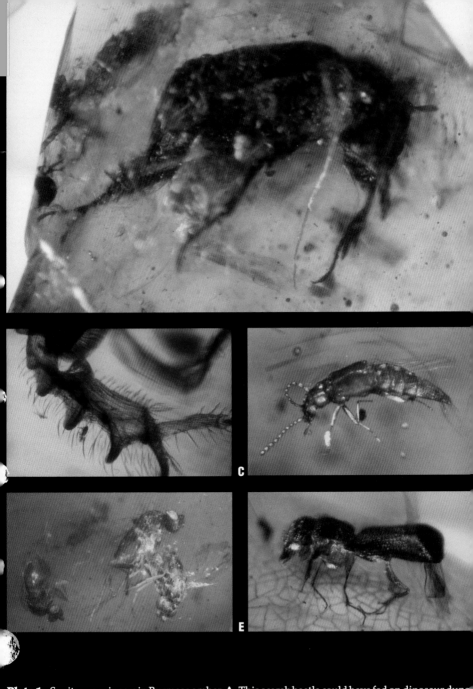

**Plate 1.** Sanitary engineers in Burmese amber. **A.** This scarab beetle could have fed on dinosaur dung. **B.** Foreleg of the above scarab resembles that of a modern dung beetle. **C.** Staphalinids (rove beetles) could have preyed on insects developing in dinosaur dung or carcasses. **D.** These or related phoric flies could h̶e̶ a fed on dinosaur remains. **E.** A small scarab with unknown habits.

**Plate 2.** Plant pests and vectors. **A** and **B.** These planthoppers in Burmese amber could have transmitted viruses to dinosaur food plants. **C.** A group of aphids in Canadian amber indicates that large populations of these potential vectors were present in the Late Cretaceous.

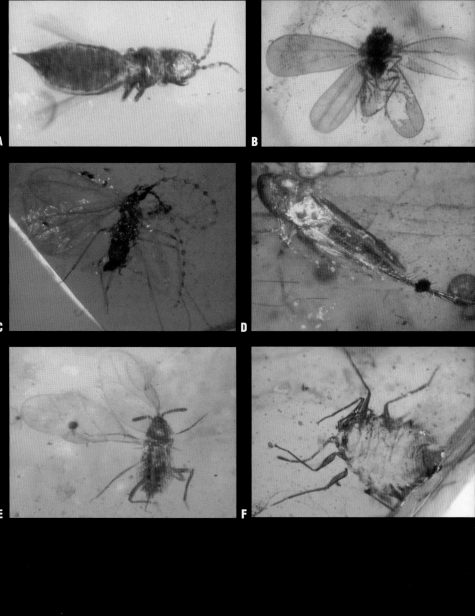

**Plate 3.** Plant pests and vectors in Burmese (A, B, D, E, F) and Canadian (C) amber. **A.** This thrips (Ectinothripidae) could have transmitted bacterial, viral, and fungal pathogens to dinosaur food plants. **B.** Whiteflies are another potential vector of plant viruses. **C.** This gall gnat belongs to a group (Asynaptini) whose descendants develop in conifer cones. **D.** Leafhoppers were another potential vector of plant viruses. **E.** This aphid (*Burmitaphis prolatum*) could have transmitted plant viruses.[43] **F.** Coccids could have killed plants with their toxic secretions as well as provided honeydew for biting insects.

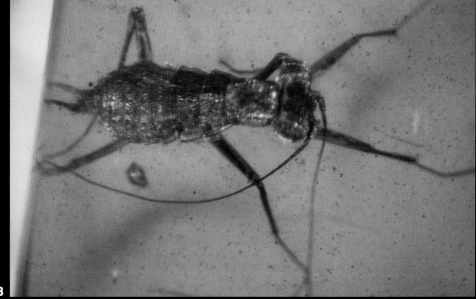

**Plate 4.** Various orthopterans in Burmese amber, such as this cricket (A) and long horned grass-hopper (B), would have competed with dinosaurs for plant resources, as well as providing food for them.

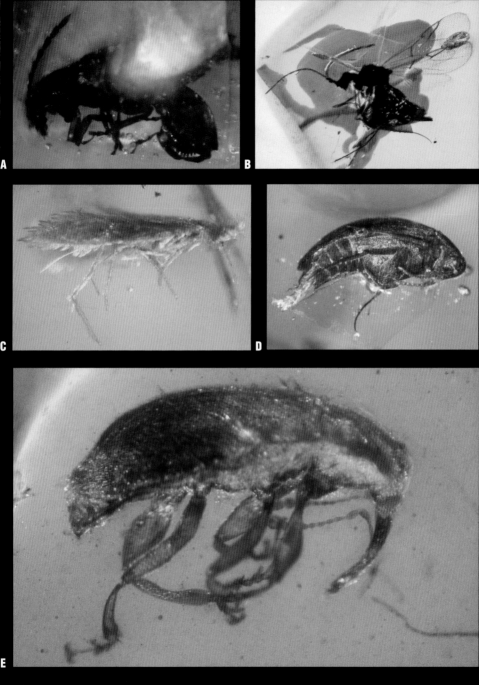

**Plate 5.** Food competitors in Canadian (A, B) and Burmese (C, D, E) amber. **A.** This bruchid palm beetle[26] would have taken its toll of seeds. **B.** This small wasp may have developed in araucarian seeds. **C.** Larvae of this moth could have also fed on dinosaur food plants. **D.** Dinosaurs would have encountered tumbling flower beetles (Mordellidae) in fungi or plant stems. **E.** A weevil whose larvae could have fed on dinosaur food plants.[41]

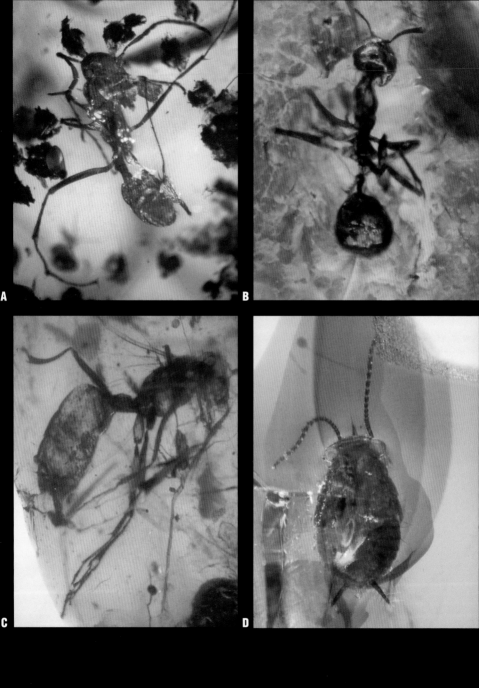

**Plate 6.** Insects as food in Canadian (B, D) and Burmese (A, C) amber. **A, B.** Workers and immatures of these ants would have provided meals for dinosaur hatchlings. **C.** This winged ant indicates a high level of social development. **D.** Cockroaches would have served as food for dinosaurs as well as carried internal parasites.

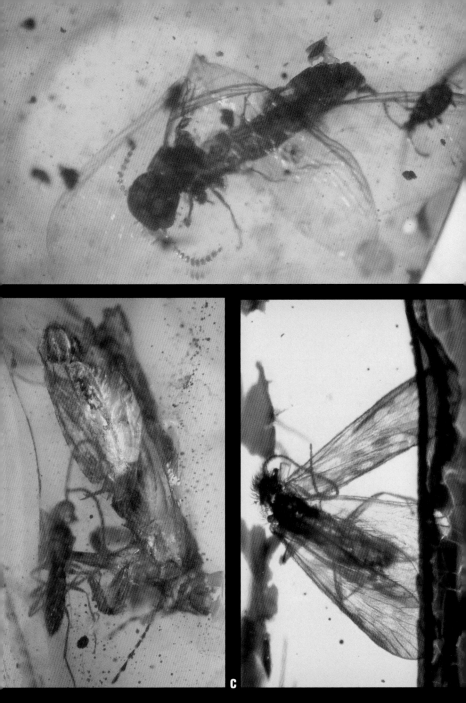

**Plate 7.** Insects as food in Burmese (A, B) and Canadian (C) amber. **A.** All termites would have provided a food source for young dinosaurs. **B.** Congregations of these pygmy grasshoppers could have been quite nutritious. **C.** This adult caddis fly, as well as its larvae, were available food sources.

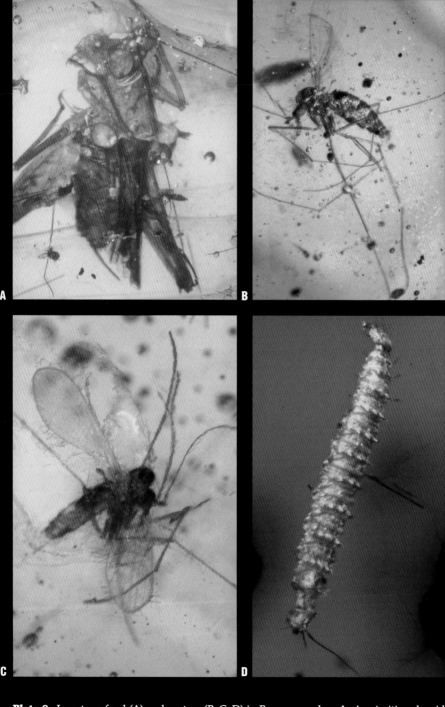

**Plate 8.** Insects as food (A) and vectors (B, C, D) in Burmese amber. **A.** A primitive elcanid grasshopper.[50] **B.** A sand fly vector of leishmania. **C.** This sand fly contains reptilian blood cells infected with leishmania in its gut.[255, 260] **D.** A rare sand fly larva associated with a club mushroom. The adult sand fly could have fed on dinosaurs.[181]

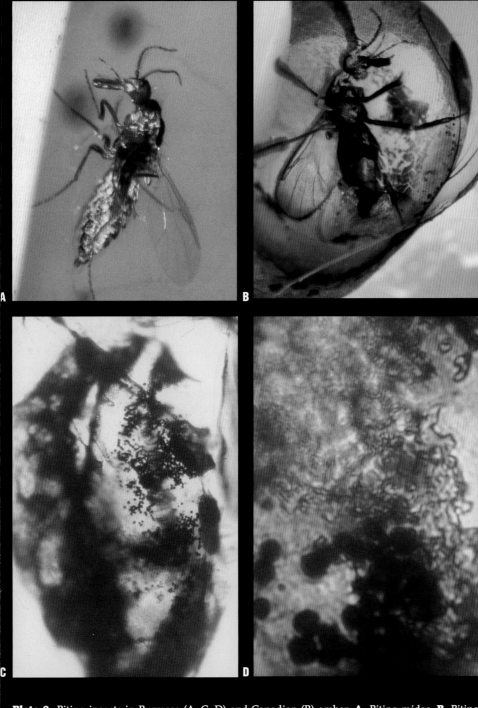

**Plate 9.** Biting insects in Burmese (A, C, D) and Canadian (B) amber. **A.** Biting midge. **B.** Biting midge. **C.** Polyhedra of a cytoplasmic polyhedrosis virus in the midgut wall of a biting midge.[172] **D.** Polyhedra of a cytoplasmic polyhedrosis virus, together with flagellates, in a biting midge. [172]

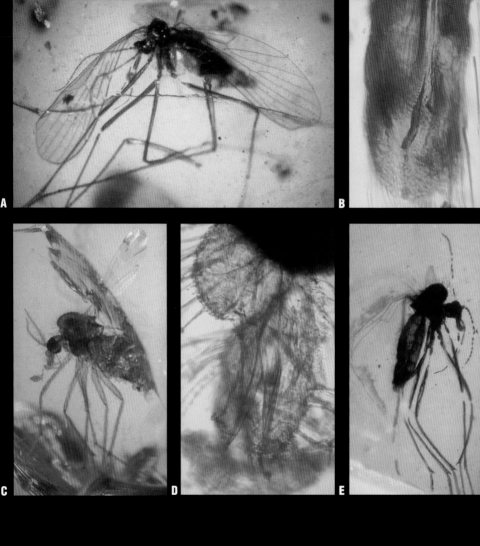

**Plate 10.** Biting flies in Burmese amber. **A.** This tanyderid fly [46, 47] could have fed on dinosaurs. **B.** Serrated mandibles on the tanyderid could have easily cut through dinosaur skin. **C.** A chironomid with biting mouthparts. **D.** Proboscis of the chironomid in C showing well-developed mandibles. **E.** Another lineage of biting chironomids.

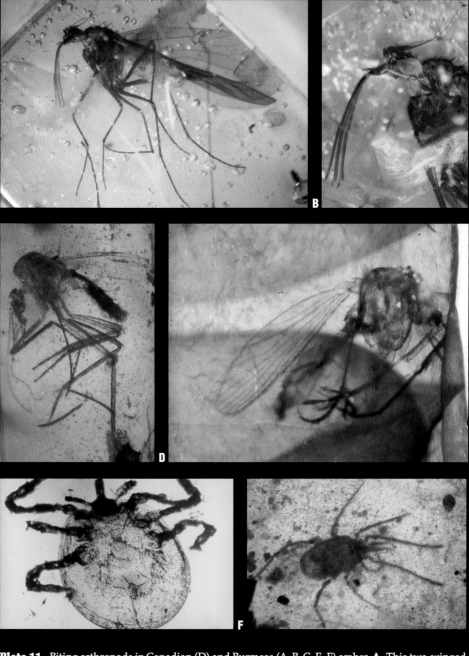

**Plate 11.** Biting arthropods in Canadian (D) and Burmese (A, B, C, E, F) amber. **A.** This two-winged scorpion fly resembles a large-beaked crane fly. **B.** Serrations along the edge of the scorpion fly's mandibles indicate that it could have easily cut through skin and scales. **C.** A biting corethrellid fly.[338] **D**. This lineage of mosquitoes[191] could have been transmitting viruses, malarial pathogens, and filarial nematodes to dinosaurs. **E.** This hard tick[45] could have fed on and transmitted pathogens to dinosaurs. **F.** A parasitic mite.

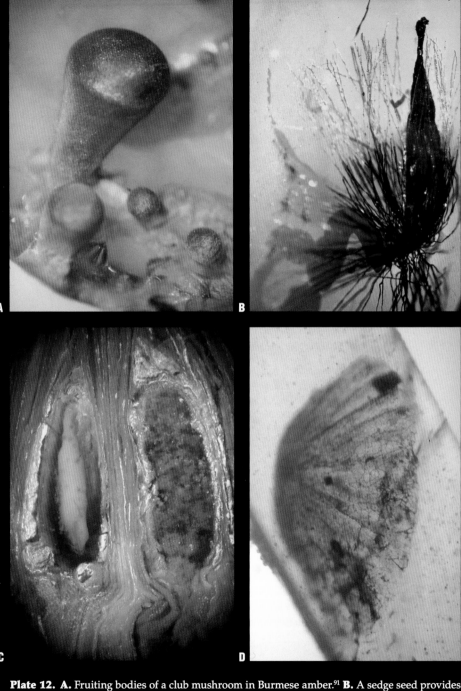

**Plate 12.** **A.** Fruiting bodies of a club mushroom in Burmese amber.[91] **B.** A sedge seed provides evidence of a food source for ceratopsin dinosaurs at the Canadian amber site. **C.** Insect frass filling a seed cavity (on the right) of a permineralized araucarian cone (*Araucaria mirabilis*) shows that insects were using that food resource even in the Jurassic. **D.** This Burmese amber mushroom probably grew on the trunk of an araucarian tree.[345]

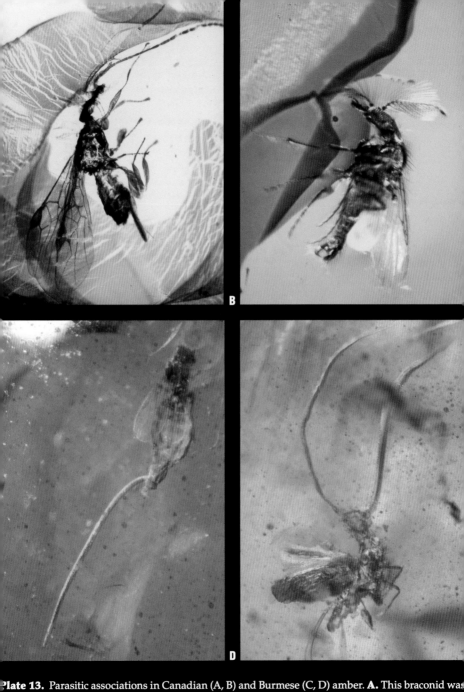

**Plate 13.** Parasitic associations in Canadian (A, B) and Burmese (C, D) amber. **A.** This braconid was probably carrying polydna viruses that protect its larvae from defense reactions of the insect host. **B.** A trombiculid mite feeding on a biting midge. **C.** A nematode parasite emerging from a biting

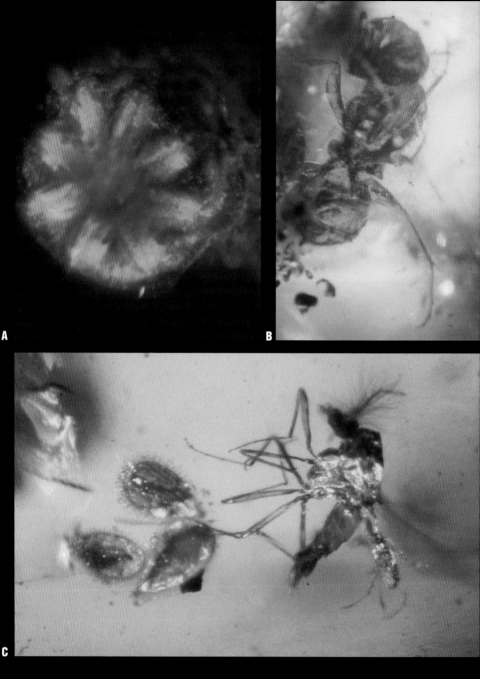

**Plate 14.** Rare inclusions in Burmese amber. **A.** This small staminate flower (*Palaeoanthella huangii*)[18] shows insect-feeding damage. **B.** This primitive bee (*Melittosphex burmensis*)[44] could have pollinated *Palaeoanthella*. **C.** A midge with a parasitic mite on its back adjacent to three seeds covered with glandular hairs. The sticky hairs indicate that animals distributed these seeds.

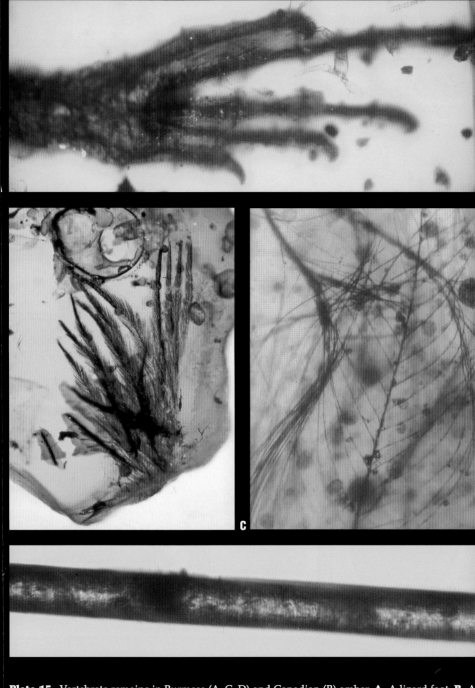

**Plate 15.** Vertebrate remains in Burmese (A, C, D) and Canadian (B) amber. **A.** A lizard foot. **B.** A group of feathers. **C.** Experts are stumped attempting to identify the bird (or dinosaur) that bore these feathers. **D.** This strand of mammalian hair still retains its scale pattern.

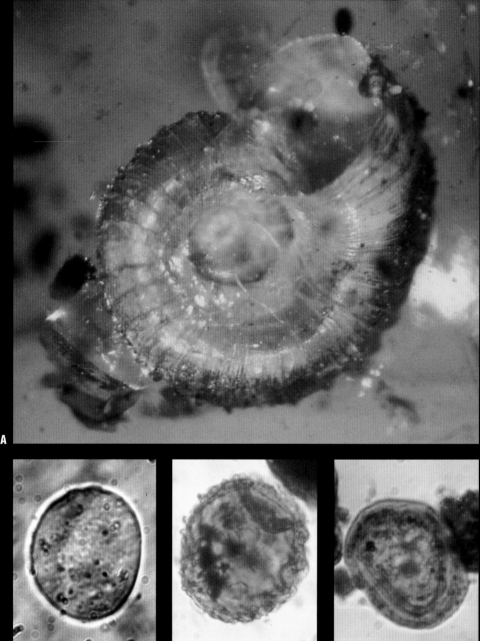

**Plate 16. A.** Gastropods like this pygmy snail of the family Punctidae in Burmese amber could have served as intermediate hosts of internal parasites of dinosaurs. **B–D**: Parasite stages extracted from a dinosaur coprolite collected at the Bernissart Iguanodon Quarry in Belgium.[135] **B.** A cyst of the protozoan *Entamoebites* that causes dysentery in reptiles, birds, and mammals today. **C.** An ascarid egg with the outer cortical layer still attached. **D.** An ascarid egg that lost its outer cortical layer.

# 19.

## Parasitic Worms

*An ailing ankylosaur fell behind the herd, obviously in pain. Masses of roundworms in the stomach had obstructed the normal flow of food through the alimentary tract. Many of these ascarid worms had reached more than a foot in length, and were depositing millions of minute eggs that then passed out the partially blocked intestine in the stool. Cockroaches and other insects gathered to feed on the broken-down plant material in the feces and were accidentally ingesting the eggs. When they in turn were eaten by young dinosaurs, the infection cycle would be continued.*

*Along with the ascarid eggs in the stool were eggs of trematodes, a type of fluke that lived in the gall bladder of the dinosaur. Eggs of these parasites were washed by the rains into a nearby stream where they hatched into minute larvae that entered snails. The trematodes would have to undergo several additional developmental stages before finding their way back into a dinosaur.*

*A mosquito landed on the vulnerable underside of the ankylosaur and began taking blood and ingesting microscopic filarial nematodes that were circulating through the capillaries. These nematodes would pass through several stages in the insect before being ready to finish their life cycle in another dinosaur. Once inside a dinosaur, the filarial nematodes searched out specific tissues to commence growing, often reaching several feet in length. All of these worm parasites made life miserable for this and many other dinosaurs.*

Parasitic worms, also known as helminths, are widespread in all vertebrate groups today. The eggs of these parasites have been recovered from prehistoric and fossilized coprolites of many animals, including dinosaurs.[135] So we already know that dinosaurs harbored both stomach worms (ascarids) and trematodes in their gut. And if nematodes and trematodes were present in the Cretaceous, we can assume that tapeworms and spiny-headed worms were also there. What, then, was the effect of these parasites?

There are many groups of nematodes besides ascarids that had the potential to infect dinosaurs, including types that live in the alimentary tract, blood, and internal tissues.[173,242,243] Those carried by arthropods have already been discussed under the specific vectors, but many others are passed on when a variety of different invertebrates are ingested.

Primitive lungworms (Rhabdiasidae) probably parasitized dinosaurs since they are widespread in lizards today. Species of *Pneumonema* reach their host by first burrowing into snails, then waiting until the mollusks are eaten by lizards such as blue tongues. Once free in the reptile's intestine, these roundworms burrow through the gut wall, enter the body cavity, and migrate into the lungs. Nematode populations can build up to the point where they block air passages and kill the hosts.[173] Other nematodes that possibly parasitized the small intestine of dinosaurs were various types of strongyle worms (Strongylidae). With huge mouths adapted to grasping the intestinal wall of snakes and lizards, ancestors of *Kalicephalus* would have been a prime candidate for infecting these giant reptiles (fig. 26). To reach a host, the parasites first penetrate into soft-bodied soil invertebrates, like snails or insect larvae, which are then eaten by snakes and lizards. Snails from Cretaceous amber (color plate 16A; fig. 27) may have been carrying the early stages of lungworms or strongyle worms (fig. 28).

Perhaps certain other strongyles caused "creeping disease" in dinosaurs, much as they do to crocodiles today. These nematodes, whose full life cycle has not been discovered, end up in the integument of their reptilian hosts and form "trails" under the skin as they tunnel beneath the scales and deposit their eggs.[242]

FIGURE 26. Strongyle nematodes like this kalicephalid removed from a snake could have clung to the gut walls of dinosaurs with their huge mouths and pumped up host fluids with their muscular esophagus. To reach their reptilian hosts, these parasites penetrate into soft-bodied invertebrates, like snails or insect larvae.

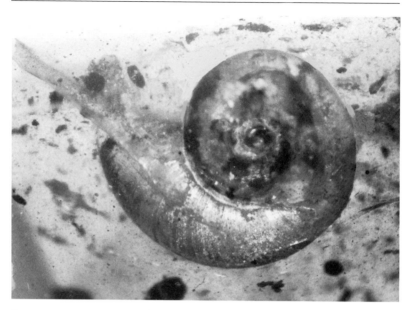

FIGURE 27. Snails, possibly including this Burmese amber specimen, are capable of carrying various worm parasites of vertebrates.

Scores of spirurid nematodes are known to parasitize reptiles and birds. In order to reach feeding sites in the stomach of vertebrates, their eggs are eaten by dung- frequenting insects such as beetles and cockroaches.[244–246] When these insects are consumed by a lizard, bird, or mammal, the worms leave their cysts (fig. 29) and establish themselves in the vertebrate's alimentary tract. Dinosaurs probably became infected in much the same way since cockroaches were plentiful back then. Heavy infections then subsequently contributed to an early death, such as occurs in our parasitized domestic swine. Eyeworms, types of spirurids that colonize the orbits of their victims, also use cockroaches as intermediate hosts. Large numbers of eyeworms in the orbits of dinosaurs might have blurred their vision (fig. 30), just as they do to birds today.[173]

Filarid nematodes, which occur in the blood and tissues of birds, mammals, and reptiles, certainly parasitized at least some

FIGURE 28. One of several juvenile nematodes found adjacent to the snail in color plate 16A. Several types of nematodes are associated with present-day snails, including those that parasitize the lungs of vertebrates.

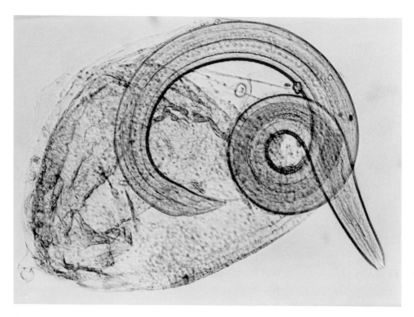

FIGURE 29. A juvenile spirurid nematode emerging from its protective cyst in the body cavity of a present-day scarab beetle. Dinosaurs could have acquired these worms by eating infected cockroaches or dung beetles.

FIGURE 30. Eyeworms transmitted by infected cockroaches could have blurred the vision of dinosaurs.

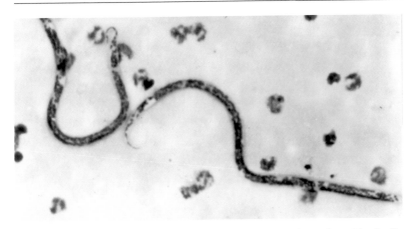

FIGURE 31. Two microfilaria of a filarid nematode, together with vertebrate blood cells, circulate in the host's bloodstream until picked up by a bloodsucking arthropod.

of the dinosaurs. Bloodsucking arthropods transport a wide variety of these roundworms, and the miniscule size of the juvenile nematodes (microfilaria) (fig. 31) is a marvelous adaptation that makes it possible for them to pass through the mouthparts of the vector when blood is consumed.

Because of their large size, stomach worms (ascarids) are readily noticed when they appear in stool samples (fig. 32). They can easily reach a foot in length, and when abundant, may cause intestinal blockage and death. Highly resistant eggs that can withstand heat and drought are the key to their success. The eggs can even become airborne, so just taking a deep breath can begin the infection process. Stomach worms currently parasitize birds and reptiles, including lizards, chameleons, and monitors,[247] and there is evidence that chameleons can acquire the parasites just by eating contaminated mosquitoes.[248] A few lizards acquire the worms by ingesting ants,[249] and if lizards are scarce, the nematodes are able to complete their development in the ants. Intermediate hosts for bird ascarids include crickets, beetles and earwigs,[173] all groups that occurred throughout the Cretaceous. We know that dinosaurs were parasitized by stomach worms,[135] but what would their symptoms have been (color plates 16C, 16D)?

Figure 32. Ascarid nematodes that lived in the stomachs of dinosaurs may have caused considerable discomfort to their hosts[135] (color plate 16C, 16D). Various orthopterans, beetles, and earwigs could have served as intermediate hosts.

These parasites can cause nutritional disorders as well as alimentary tract and organ blockage in reptiles.[203] Today, nearly 1.5 billion humans are infected with ascarid worms, some 340 million suffer serious side effects, and about 100,000 die every year.[3] Other nematode groups that were probably associated with dinosaurs were pinworms in the large intestine, guinea worms in the tissues, and capillarids and camallanids in the alimentary tract.

Did dinosaurs have tapeworms? It is quite likely since at least 70 of these parasites infect lizards and snakes and many more are found in mammals and birds.[250] Most tapeworms of reptiles belong to a primitive Order (Proteocephalidea), which includes species that live in the small intestine of monitor lizards.[203] Related genera also parasitize a wide variety of lizards, sometimes even invading their body cavities.

Many reptilian tapeworms require several hosts to complete their development. The first is often a copepod, a type of crustacean wherein development can proceed only so far. These are eaten by a second host, normally a frog, fish, or reptile, and maturation continues through the next stage. Tapeworms reach adulthood when that animal is eaten by a third definitive (final) host. This process can take years and the tapeworms often become agitated while waiting for this last host, causing skin lesions and hemorrhages in their second victim.[203] Both intermediate and final hosts could have been dinosaurs. Some lizard tapeworms use darkling beetles as the first host.[251] Hungry anoles end up with more than just a meal if they are unlucky enough to select a parasitized beetle. Tapeworms rarely kill their hosts, but are known to block organs and ducts, produce festering wounds from their suckers or hooks, and damage internal organs, resulting in a general debilitation.

Could dinosaurs have suffered from flatworms (trematodes)? The discovery of a trematode egg in a dinosaur coprolite from an iguanodon quarry in Belgium shows that indeed they did.[135] With over 40,000 extant species of trematodes, many flatworms probably parasitized dinosaurs (fig. 33). These parasites are able to develop in and cause injury to almost every major vertebrate organ, including the alimentary and genital systems, gall bladder, urinary bladder, and blood.[250] Each egg develops into a miracidium, which in most species enters a snail and continues through several developmental stages before ending up as an adult in a vertebrate.[203,250] Thus, any dinosaur that intentionally or accidentally ingested an infected snail then acquired trematodes. Most infections are asymptomatic in reptiles and only cause problems when the hosts are under severe stress.

Another primitive parasite group that the dinosaurs could have harbored were spiny-headed worms of the phylum Acanthocephala. These parasites infect a variety of birds, mammals, and reptiles, especially turtles and snakes.[203] The adult worms can do considerable damage burrowing into the gut wall with their hook-adorned heads. They gain entrance to the vertebrate

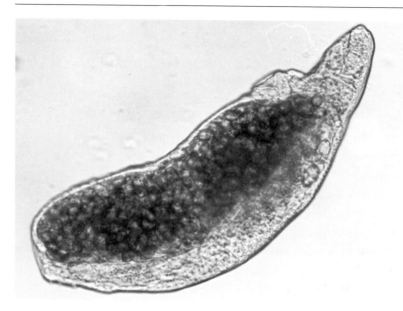

FIGURE 33. A trematode adult filled with eggs is little more than a reproductive sac. Those parasitizing dinosaurs[135] could have been acquired from eating infected snails.

by first entering the body cavity of an invertebrate that ingested their eggs. Development proceeds to the acanthella stage in the invertebrate, and when eaten, continues to the adult stage in the final vertebrate host. Besides insects, snails and crustaceans also can serve as invertebrate hosts.

If having parasitic roundworms wasn't enough, dinosaurs were impacted indirectly by other types of parasites. Chapter 7 discussed how nematodes vectored by longhorn beetles kill pines today and could have destroyed dinosaur food plants in the past. Many other types of plant parasitic nematodes develop in roots, stems, and leaves of a wide range of plants, and many of these certainly infested dinosaur food plants in the Cretaceous, impacting quality and distribution.

# 20.

## The Discovery of Cretaceous Diseases

*Just before twilight, a herd of large sauropods grazed sedately on a stand of kauri trees. With their long necks, the mature adults could easily reach up into the understory and pull down whole branches laden with succulent new foliage. The forest was filled with the sounds of eating: the thrashing of large animals through the underbrush, the cracking and rending of branches torn from the trunks, and the crashing thumps of these giants. Other, more subtle noises emanated from the herd, the rumbling from stomachs as the consumed vegetation was ground by gizzard stones and fermented in ballooning bellies, and some contented vocalization between members of the herd. The air was scented by the pungent odor of resin oozing from the damaged araucarians. Suddenly masses of biting flies rising from their resting places on the bark of the trees engulfed the herd. Included among them were female sand flies looking for a source of blood to ensure egg development.*

*After settling on a sauropod, one female fed repeatedly, acquiring not only blood for her eggs but also something much more significant—pathogenic microorganisms. The insect was interrupted in the middle of what was to be its final meal; perhaps a therapod appeared and the startled sauropod responded by scraping against some vegetation, forcing the sand fly to quickly flee. Her escape route led to the trunk of a resin-coated araucarian tree. Once in the sticky trap, she couldn't pull free. The fly's straining, desperate movements attracted the attention of a small predator patrolling the bark, who nipped open a miniscule hole in the end of her abdomen, deftly pulled out the reproductive system, and devoured the protein-rich eggs. Some of the gut contents of the entrapped insect spilled out onto the fresh resin as life ebbed away. She lay on one side in a drop of spilled blood, disemboweled, head and mouthparts clearly visible, wings outstretched—a sudden and unexpected victim of double jeopardy. An additional resin flow entombed the small female fly (color plate 8C).*

Long before our search for Cretaceous diseases began, events as described above took place in a Burmese forest 100 million years ago. The last few seconds in the life of this fly were extremely important, for they would set the stage for the very first discovery of a Cretaceous insect-vectored disease.

One of many theories presented on the demise of the dinosaurs states that they succumbed to diseases.[25] While intriguing, there was no evidence that terrestrial vertebrate pathogens existed in the Mesozoic—until recently. This is the story of how they were discovered.

We have been interested in the history of infectious organisms for decades. Conversation after conversation has centered on how to find pathogens in the fossil record, when and how did a disease arise, and where did the ancestor of these organisms first appear. Our belief has always been that amber represents the most ideal place to look for answers because of its remarkable preservative qualities. Delicate structures such as bacteria, pollen, and even cells with nuclei and mitochondria have been preserved. So if one believes that agents of disease can indeed be preserved in amber, then what kind of pathogens should one expect to find?

The answer is an arthropod-borne pathogen because amber is the best-known repository of fossil arthropods. So, insects in amber became the starting point in our search for Cretaceous diseases. This quest is where the role of a scientist becomes much like that of a detective: searching for suspects, looking for clues, applying deductive reasoning, and hoping for a bit of luck.

In the preliminary investigation, a scientist would screen arthropods that are already recognized as vectors of disease. For example, we know that mosquitoes carry malaria, yellow fever, and West Nile virus, tick bites are responsible for Lyme disease, and blackflies transmit nematodes causing river blindness. To increase the chances of finding a pathogen, the vectors one examines should be abundant, the organism must be detectable with the light microscope, and the infection should be confined to a specific location in the arthropod. Applying these parameters started the game.

Some 100 million years after becoming entrapped in resin, a remarkable treasure found its way to our laboratory, one of several biting insects recovered in Burmese amber that met our criteria for possibly containing pathogens. When the piece arrived, I carefully placed the specimen, which was a sand fly, on a microscope slide and began peering through the eyepieces, hoping but not really optimistic. Such painstaking and frustrating work—looking for evidence of disease in minute fossil bloodsucking insects! What were the chances of finding anything like a microscopic pathogen inside an insect less than a millimeter in length? Searching for something so small, something measured in microns, in a specimen unprepared by conventional fixation or unenhanced by diagnostic stains was daunting, tedious work. Then you still had to consider whether the microorganism would even be recognizable.

The odds against such a discovery appear astronomical. First, a sand fly must have fed on an infected vertebrate, ingested pathogens along with blood, and then end up being preserved in amber. Furthermore, this particular piece would have to be one of the few dug out of the ground, polished, and selected among all others to be shipped out of Burma. And finally, after passing through the hands of several dealers and collectors, the specimen would eventually have to end up under my microscope.

To make matters even more complicated, the very act of examining a fossil fly for internal pathogens is beset with difficulties. To begin with, the fly has to be fairly transparent in order to examine inside the body cavity. In most cases, the integument remains opaque or only translucent, making it impossible to obtain a clear view of the internal structures. Even if the specimen is reasonably transparent, small organisms like pathogens can stick together, appearing as dark globs, or simply be masked by the similarly colored background. The ability to distinguish pathogens in amber is only possible if they were particularly well fixed by naturally occurring chemicals in the resin, which is a rare event.[252,253]

That first hopeful glance through the microscope revealed that the abdomen was partially filled with a dark substance—could that really be blood? I thought maybe, just maybe. Switching to medium power, the surprisingly clear head region sprang into view, exposing the proboscis. In the middle of this feeding tube there was a spherical object, only 4 microns in diameter, with a dark body inside. At still greater magnification, the specimen revealed several more spherical objects, one in the center and others toward the edge of the proboscis (fig. 34). What were they? I began snapping photos. Astonishingly, the pictures revealed a short, filiform structure emerging from one of the spherical objects—a flagellum! This was exciting evidence that we had found a protozoan and it could be a pathogen.

The next day was spent reexamining the specimen. Closer inspection told us the piece had to be repolished in order to see the structures better. This procedure is beset with danger because the risk of damaging the specimen increases with every polishing. The evidence already showed that this fly was extremely valuable scientifically. After several nerve-racking hours spent tediously hand polishing that small gem, we could now focus attention on the contents of the abdomen that did in fact contain a partially digested blood meal. The fortuitous removal of the ovaries by the predator enabled us to examine the midgut, which was packed with numerous long coiled bodies orientated in a myriad of positions. They were so twisted and intertwined that it was difficult to find one that was not covered by at least a portion of another, but they were definitely flagellated protozoa (fig. 35). Inside their bodies could be seen the large dark nucleus and

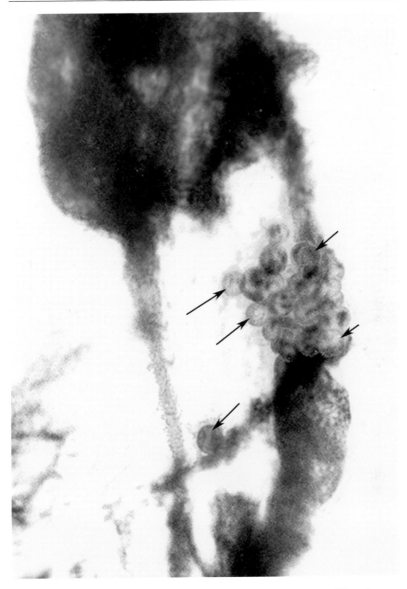

FIGURE 34. A group of amastigotes (arrows) of *Paleoleishmania proterus*[254] in the proboscis of a Burmese amber sand fly, *Palaeomyia burmitis*.

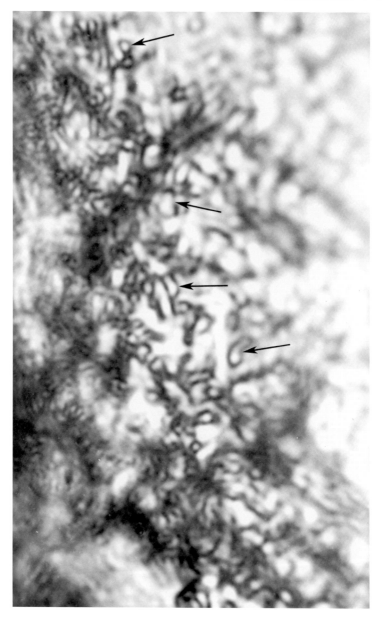

FIGURE 35. Masses of promastigotes (arrows) of *Paleoleishmania proterus*[254] in the midgut of a Burmese amber sand fly, *Palaeomyia burmitis*.

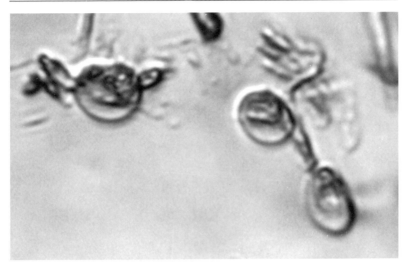

FIGURE 36. Short, oval procyclic promastigotes of *Paleoleishmania proterus*[254] associated with a Burmese amber sand fly, *Palaeomyia burmitis*.

smaller kinetoplast. Ancient trypanosomatids seen for the first time by man—the first evidence of vector-borne diseases. An amazing discovery!

Of course, these findings had to be validated by making comparisons with the various developmental stages of present-day trypanosomatids in sand flies. After reading numerous scientific papers, it became obvious that the organisms most closely related to the fossils were a group of trypanosomatids known as *Leishmania*. Could *Leishmania* have existed 100 mya? When present-day sand flies feed on a vertebrate infected with *Leishmania*, they acquire small, round, non-flagellated stages called amastigotes, which are formed in the tissues of the vertebrate. These amazingly resemble the round bodies first seen in the proboscis of the fossil. In the insect, the amastigotes then elongate and develop flagella, which provide them with mobility. These stages, the promastigotes, multiply by simple division and become quite numerous within the insect's alimentary tract. Many types occur, including oval forms called procyclics (fig. 36). Eventually some of the promastigotes become elongated and convert to nectomo-

nad stages. After a period of proliferation, the final paramastig-
ote stage is formed. These short, stubby flagellated cells migrate
to the head of the fly and are transferred to a vertebrate during
the insect's next meal. Amastigotes, nectomonads, and paramastigotes were all seen in the fossil sand fly[254] (figs. 24, 34).

Convincing photographic prints of the various stages were
needed for publication, so into the darkroom we went. Night after night, intensive work finally produced a set of prints that
demonstrated all of the characters needed to align these microorganisms with the leishmanial pathogens of vertebrates.

Now we had to obtain the next bit of evidence in our case for
insect-borne disease in the Cretaceous. With the presence of vertebrate *Leishmania* parasites established, how can the host be determined? Two genera of Old World sand flies carry *Leishmania*
to mammals and reptiles today. Those in the genus *Phlebotomus*
prefer mammals and differ morphologically from members of
*Sergentomyia* that dine on reptiles. So we turned to the sand fly
for clues that would indicate the identity of the vertebrate host.
Described in an extinct genus because it differed from all extant
types,[255] the fossil nevertheless possessed some characters found
in extant species of *Sergentomyia*. One was the position of the
hairs on the tips of the abdominal segments, which are usually
erect in *Phlebotomus* and decumbent in *Sergentomyia*. Although
only one of these remained on the fossil (many hairs and scales
of insects in amber are lost as the trapped arthropods attempt to
escape), that lone hair was decumbent. Shaky evidence, but luckily, even if hairs weren't found, their sockets were still present
and decumbent hairs have elongate sockets, which was the condition of the sockets in the specimen.

Another identifying character dealt with the wing venation,
and this fly's venation definitely aligned it with the reptile feeders. This was all very interesting and while showing that the
sand fly possessed characters of present-day reptile feeders,
there still wasn't enough evidence to conclude that she fed on
Cretaceous reptiles. Perhaps, just by some wild chance, there
were some blood cells remaining from her last meal. Vertebrate

blood cells are normally broken down within four to five hours after feeding,[142] but we had already concluded from the blood volume remaining in the abdomen of the fossil that only the early stages of blood digestion had begun before death,[255] so some of the vertebrate blood cells might still be intact.

We turned back to the fossil for more answers. Where else could one expect vertebrate blood cells to occur? The abdominal midgut had already been thoroughly examined when the various stages of the flagellates were photographed. That left only the thoracic midgut, a region that is usually obscured by thick cuticle and dense flight muscles. And as expected, this area appeared to be completely opaque. A seemingly hopeless endeavor, but still worth a try. After several hours of orientating the fly in different planes, a miniscule spot of light suddenly opened a view into the thoracic midgut lumen. Illuminated in this small expanse was a group of nucleated vertebrate blood cells. Another breathtaking find!

Taking and processing several rolls of film eventually led to satisfactory prints of these enigmatic nucleated cells, and the process of identification began. Measurements were taken and the literature on vertebrate blood cell size and morphology was consulted. The nucleated blood cells in the fossil midgut lumen were oval, ranging from 10 to 15 µm in length. Mammals were eliminated as a source because most of their blood cells are discoidal and enucleated. However, oval nucleated blood cells occur in birds, amphibians, and reptiles.[256] How could we narrow the search further? Amphibians have large blood cells, ranging from 18 to 67 µm in greatest dimension (most are over 26 µm), and since there are also no cases of *Leishmania* or sand fly vectored trypanosomatids in amphibians today,[183] we could strike them off the list of suspects, leaving birds and reptiles. Both reptiles and birds have blood cells overlapping in size with those in the fossil.[256] However, there are no present-day cases of *Leishmania* or other sand fly-transmitted trypanosomatid infections in birds,[257] which eliminated them from further consideration and left only one group, the reptiles.

We know that there are numerous recorded cases of try-panosomatid infections in lizards and snakes, all of which are known or suspected of being transmitted by sand flies.[183,258,259] Also, one of the complete nucleated blood cells in the fossil insect is almost identical in shape and size to a lizard proerythrocyte.[183] All of these facts taken together, from the characters of the vector fly to the morphology and size of the blood cells, led to the verdict that the fossil sand fly had been feeding on a reptile.

In *Leishmania* infections, the amastigotes are produced in blood cells of vertebrates, where they appear as small, dark areas inside lighter areas known as parasitophorous vacuoles. The most astonishing thing is that some of the fossil blood cells actually contained these vacuoles with developing amastigotes inside (fig. 37). They bore a striking resemblance to developing *Sauroleishmania* amastigotes in the blood cells of present-day lizards.[183,260] So here were infected reptilian blood cells inside this little sand fly. The chances of finding evidence of such pathogens among the fossil remains of any vertebrate would be miniscule. That is why amber is a treasure trove with so many secrets from the past.

What an intricate case of detective work this turned out to be. The implications of this find did not fully sink in until we began to wonder what types of reptiles might have provided the meal for our incredible sand fly. What were the dominant reptiles, or should we say the dominant vertebrates at that time—none other than dinosaurs! Could this fossil sand fly have been feeding on a dinosaur? Why not? Certainly they were widespread and since reptile-seeking sand flies today prey on a variety of lizards and snakes, why wouldn't Cretaceous forms dine on dinosaurs? Suddenly those cells took on a new light: they could be the first infected dinosaur blood cells seen by man. And much more significant, might they contain agents that caused, or at least contributed to, the demise of the dinosaurs? We can't say for sure, but we know that sand flies were widespread in the Cretaceous, probably with a global distribution. With many of the continents connected, pandemics could easily have occurred. This one fos-

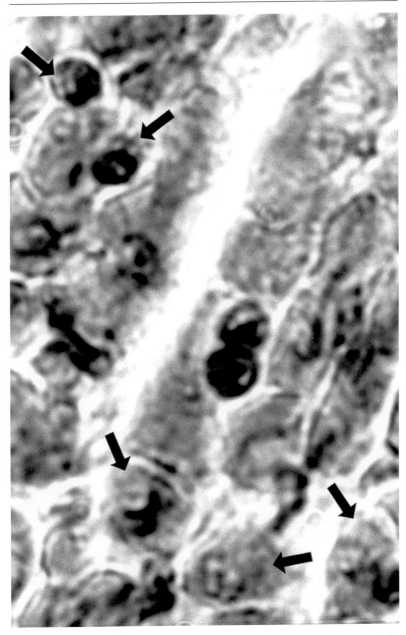

FIGURE 37. Reptilian blood cells inside the gut of a Burmese amber sand fly, *Palaeomyia burmitis*. Arrows show amastigotes developing in vacuoles within the blood cells.[260]

sil, with its immense wealth of scientific information, opened up a large view to the past, one that revealed the stark reality of infections, death, and possible extinctions.

While finding a single infected vector establishes the presence of a pathogen at a particular place and time, this alone cannot indicate the prevalence of infection. That is why we were quite excited to discover additional infected sand flies in Burmese amber obtained over the next few months. In fact, out of 21 female sand flies that have been examined, 10 were found with trypanosomatids. This discovery confirmed our original suspicion that the incidence of reptilian leishmaniasis was extremely high in that location 100 mya. The chances of finding an infected present-day sand fly in regions with natural leishmanial infections are very low. What would then be the incidence of infection based on the sand flies found so far in Burmese amber? Incredibly high! Perhaps most or all of the resident sand flies were feeding on infected vertebrates, indicating an epidemic among the vertebrate hosts, possibly including dinosaurs.

If one disease agent could be found, what about others? Our interest was whetted and we began to investigate biting midges as possible vectors of vertebrate pathogens. Our perseverance was rewarded when we discovered an ancient malarial organism in a Burmese amber ceratopogonid (color plate 9A). Inside the body cavity of the fossil fly were developing oocysts (fig. 38) and sporozoites of a *Haemoproteus*-like pathogen, all completely preserved (fig. 23). Again using morphological characters of the biting midge, we determined that it probably fed on a large vertebrate, and the predominant large vertebrates at that time were dinosaurs.[261]

While protozoa are visible with the light microscope, viruses are not. But maybe we could find some insect-pathogenic viruses with a marker visible with the light microscope, like proteinaceous polyhedral bodies that encapsulate the virions. They would establish the presence of certain virus groups. Again, persistence paid off. The midgut of a biting midge in Burmese amber contained several hundred polyhedra (color plates 9C, 9D) of

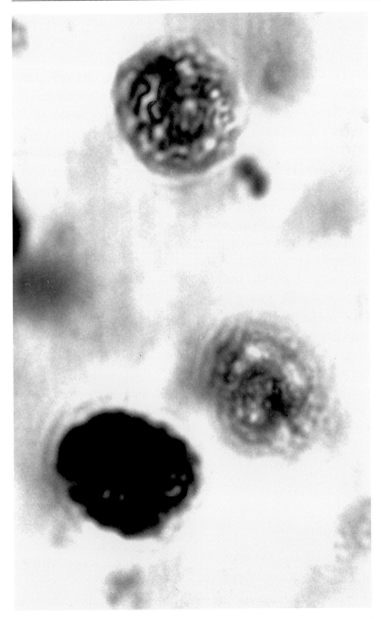

FIGURE 38. Developing oocysts of *Paleohaemoproteus burmacis* in a Burmese amber biting midge.[167]

a cytoplasmic polyhedrosis virus.[172] As occurs in this type of present-day virus infections, the polyhedra were localized in the midgut of the host. Cytoplasmic polyhedrosis viruses are members of the viral family Reoviridae, which also includes some vertebrate pathogenic arboviruses. So while arboviruses have not been recovered from fossil arthropods, the polyhedra of insect parasitic viruses that could be the precursors of some types of arboviruses have been found. Today, cytoplasmic polyhedrosis viruses infect biting midges, mosquitoes, and sand flies, all of which also transmit arboviruses to vertebrates. The viral-infected Burmese amber biting midge in all likelihood, based on various morphological features, fed on vertebrates.[172]

Looking at additional amber fossils has told us something about other insect diseases in the Cretaceous. We have found a putative nuclear polyhedrosis virus in a sand fly, a fungal pathogen of a scale insect (color plate 13D), mite parasites attached to insects (color plates 13B, 14C), and a nematode parasite of a biting midge (color plate 13C).[337] Finding a pathogenic fossil organism only tells us that it was present at a particular point in time. How long it existed before that can only be determined when older fossils are discovered. Logic tells us that many diseases were well established long before our samples appeared. Just how long remains to be seen.

# 21.

## Diseases and the Evolution of Pathogens

ONE OF THE BASIC PREMISES of this book is that Cretaceous insects transmitted pathogens that either directly or indirectly affected dinosaurs. The results were not only dinosaur disease and mortality but also the destruction of dinosaur food plants. We feel that most, if not all, present-day vector-pathogen associations were already established or arose at some point in the Cretaceous, even though different genera and species of hosts and vectors were involved.

The origins and coevolution of pathogens with their vectors and plant or animal hosts are complex. When the first dinosaurs walked the earth, they already came with pathogens carried over from their ancestors. While new vector-pathogen associations were evolving, the basic survival plan of many pathogenic microorganisms had already been established.

People are in contact with microbes from the day they are born. Each of us is a walking, breathing ecosystem carrying around thousands of organisms on and in our bodies. It is said that an individual has more bacteria in his or her mouth than there are people on the surface of the globe.[66] And unfortunately, sometimes we pick up pathogenic microbes.

The number of human lives lost from infectious diseases throughout recorded history far exceeds deaths from famines and war together. Epidemics have destroyed armies, wiped out indigenous peoples, halted human enterprises, and inflicted untold misery. And if not for drugs, insecticides, vaccinations, and our ability to understand the causes of diseases and act accordingly, our current census figures would be much lower. How many more of our species would have died from malaria and yellow fever if we had not cleared the land, drained the swamps, used tons of pesticides, and built dwellings with screened windows? Yet even with all of this, yearly mortality rates from just malaria are well over a million. What role did diseases play in regulating the distribution of animals and plants in ancient landscapes?

Vertebrate diseases are commonly lumped into two categories, contagious and noncontagious. In contagious illnesses, the pathogens are spread from one individual to another by direct or indirect contact. Touching the infected individual, breathing the same air, coming in contact with contaminated waste products, and so on all serve to spread the infectious agent. On the other hand, noncontagious diseases need to be passed on by another organism, commonly known as a vector, that inoculates the pathogen into the host. Just coming into contact with sick individuals will not result in infection. We have shown that vector-borne pathogens existed in the Cretaceous. However, evidence of contagious diseases has not been discovered, although they were certainly present back then. Especially lethal representatives of both types of diseases have been considered for use as biological weapons.[262]

In general, epidemics causing widespread mortality originate in two ways. Either a pathogen already present in the ecosystem mutates into a more lethal strain or an already existing pathogen makes contact with a new, exotic host. An example of the former would be the bird flu virus that has developed strains capable of infecting humans. Instances of the latter were the introduction of smallpox and measles into the New World and their widespread destruction of Amerindian populations. Perhaps the most dangerous situations are those that combine both features, that is, a newly introduced, actively mutating pathogen.

So when and how did pathogens arise? Scientists believe that the precursors of life arose in primordial oceanic ooze as simple organic molecules that polymerized into complex amino acids, the building blocks of life. These were incorporated into abiotic protocells, and when nucleic acids were manufactured in that generative cauldron, true cells were formed. Those very first pre-life forms were probably similar to two unusual pathogens. The simplest of these are the prions, which are composed only of a single protein. Viroids are more complex and are made up of a short chain of naked RNA. Their elementary structure suggests that these were the most ancient pathogens.[263,264]

Following along with this line of thinking would mean that the first diseases probably occurred in bacteria. These prokaryotes, cells without nuclei and most organelles, first appeared as fossils 3.5 billion years ago. They then had 2 billion more years to experiment with diversification and develop complex metabolic pathways. During this interval, viral pathogens such as bacteriophages had sufficient opportunities to arise, invade, and coevolve with bacteria, thus establishing one of the earliest disease associations.

By 1.5 billion years ago, fossil eukaryotes—organisms with well-defined nuclei and organelles—arrived on the scene. There is good evidence that two of these organelles, mitochondria and chloroplasts, represent small symbiotic bacteria that pushed their way into larger eukaryotic cells and remained. This would be extremely significant if it meant that the invasions of endosymbionts were responsible for the critical transition from prokaryotic to eukaryotic life. After all, in order to function, the majority of all life on earth depends on mitochondria, and the plants that form the basis of our food chain require chloroplasts! There is data showing that mitochondria from higher plants are so different from that of all other eukaryotes that their invasion had to have taken place on a separate occasion. This means that at least three different symbionts were able to gain permanent entry to primitive cells. Furthermore, the first eukaryotes were probably protists, and many of these are known to be hosts to endosymbiotic bacteria as well as to other smaller protists. If one accepts that these mutualistic associations were present, than why not parasitic ones as well? The same invasive mechanisms could easily have worked for both beneficial and detrimental symbionts. Most pathogenic microorganisms, as well as metazoan parasites, evolved from free-living ancestors that lived as saprophytes, predators, or ectoparasites of other organisms.

Over eons, viruses, bacteria, fungi, protozoa, and nematodes formed endosymbiotic associations with higher life forms, including vertebrates. As pathogens experimented using the vertebrate body for food and shelter, they elicited various responses in the host, the most drastic being death. Each infected animal developed ways to combat these invaders. Just when immune responses appeared in eukaryotes is unknown, but at some point vertebrates developed an arsenal of general and specific defensive weapons to use against foreign bodies. The first line in a general defense is a protective integument to keep the organisms out. However if this barrier is broken or entry occurs through the digestive or respiratory systems, blood cells then may engulf and destroy the parasites. Protective proteins, including complement and interferon, also battle the intruders. When this doesn't work, a more specific battle is mounted and an immune response triggered by antigens on the surface of the invaders stimulates the production of antibodies. If the host survives that first contact, it has an acquired immunity towards that pathogen, which may be complete or only partial.[265]

When the immune system is overwhelmed or compromised from dealing with too many infections, the host may perish. Whether a vertebrate can survive an infection depends on the number of parasites introduced (the infective dose), the immune system response (whether there was previous exposure), and environmental conditions.[266] Throughout recorded history, deadly epidemics in humans were often caused by pathogens not previously experienced. These "novel" or "emerging" pathogens are usually mutated strains of existing organisms or the transfer of disease-causing agents from reservoir animal hosts (zoonosis).

What kind of resistance to disease did dinosaurs have, and is it possible that their immune systems were not prepared for certain pathogens? Were dinosaurs unable to produce the antibodies crucial in defending them against protozoan and viral pathogens?[267] There is some evidence of pathogens and parasites associated with dinosaur fossils. Aside from the presence of gastrointestinal parasites in dinosaur coprolites[135] (color plates 16B, 16C, 16D), some observations might pertain to infectious microbes in dinosaur blood vessels. In a microscopic preparation of a sauropod bone from Wyoming, Moodie[268] observed ovoid bodies around the periphery of the vascular spaces. While some of these bodies resembled reptilian blood cells, others were irregular and appeared to be agglutinated corpuscles. Roy Moodie mentioned that similar small, round bodies had also been observed in the Haversian canals of an iguanodon. Is it possible that some of these were infectious microorganisms? A more recent study showed putative blood cells in a vessel of a *T. rex* containing diffuse gray bodies, which may represent developmental sites for malaria or leishmanial parasites.[269]

Today, most vector-borne infectious diseases occur in tropical or subtropical climates, which suggests that pathogens, hosts, and vectors originated under similar conditions. During the Cretaceous, when we propose that vectors and pathogens were coevolving, the climate was predominately tropical to subtropical.[180]

### Viruses

As mentioned earlier, viruses are assumed to be quite ancient and their first hosts were likely bacteria. There are several hypotheses on virus origins, one of which states that they extend back to the beginning of life some 3.8 billion years ago. This seems plausible and would give viruses an extended period to invade higher life forms as they appeared on the scene. Today some 1,950 species of viruses have been described from a wide range of plants and animals.[272] The origins of various lines have been estimated using molecular techniques. While those methods can estimate when various lineages arose in the past, it is still good to have fossils to confirm the time and place. As would be expected, the fossil record of viruses is virtually nonexistent, except for putative cytoplasmic and nuclear polyhedrosis viruses in adult biting flies in Burmese amber.[172] Indirect evidence of polydna viruses is found with fossil braconid wasps, whose present-day counterparts utilize these viruses to suppress host immune reactions (color plate 13A). Discovery of these insect viruses provided evidence to support the hypothesis that some arthropod-borne viruses (arboviruses) evolved first as pathogens of bloodsucking flies. This seems plausible since several groups of viruses such as the Togaviridae, Reoviridae, Poxviridae, and Rhabdoviridae have members that can infect both insects and vertebrates.[271] This dual-host biology is also seen with viruses that infect both insects and plants and illustrates the wide cross-phyla host range of many viruses. It is likely that not only arboviruses but also other contagious viruses infected vertebrates throughout the Cretaceous. Whether they caused disease in dinosaurs cannot be said, but we know that the vectors that transmit them today were also present back then. Many of the insect groups that transmit plant viruses today were also present at that time, and while viruses from angiosperms abound, very few have been observed in other plant groups. Plant viruses were certainly important in defining community structure then as they are now.

## Bacteria

We have already spoken about the early evolution of the ubiquitous bacteria, of which some 10,000 species have been described so far. Fossils demonstrate that they began evolving in the oceans 3.5 billion years ago. Our knowledge of the planet's prokaryotes is extremely limited even though they are so numerous that only a milliliter of water typically contains up to a million bacteria. Being so widespread, it is not surprising that some are transferred from the mouthparts of horseflies, lice, ticks, and sand flies to vertebrates. We've discussed bacteria carried by arthropods, but other types known as facultative pathogens may have been equally important. These include the cosmopolitan pseudomonads and serratias that enter wounds and spread throughout the circulatory system, cleverly avoiding immune responses.[234] If dinosaurs had been contaminated via wounds, resulting infections could have been fatal, much as they are in lizards killed by *Serratia* today.[270]

A number of other bacteria, similar to those that cause strep throat, pneumonia, tuberculosis, and diarrhea in humans, are acquired by inhalation or via contaminated foods or liquids. These organisms, as well as anthrax that occurs worldwide in populations of wild and domestic vertebrates (also occasionally man) and is spread through cuts, abrasions, and inhalation, were probably circulating among Cretaceous vertebrates. Some of these communicable organisms could have presented serious threats to dinosaurs.

### Rickettsia

Rickettsias are minute rod-like organisms often considered bacteria. They occur in many locations inside the alimentary tract and tissues of a wide variety of animals. Some cause diseases in insects,[271] while others are pathogenic to both arthropod vectors and their vertebrate hosts. Symbiotic species in the wide-ranging genus *Wolbachia* play various beneficial roles in arthropods, even digesting portions of the blood meal in biting insects. The fossil record of rickettsia is non-existant and few scientists have described these very small organisms because they are difficult to culture. Two serious human rickettsial pathogens are those that cause the plague, transmitted by fleas, and typhus, carried by lice. Rickettsia certainly resided in dinosaurs, but whether they resulted in epidemics cannot be answered.

### Fungi

There are over 100,000 species of fungi[273] and they have established relationships with all life forms. At least 77,000 different plant-fungus associations occur in just the United States,[274] so if we consider those on the rest of the globe and also include the fungal pathogens of invertebrates and vertebrates, as well as those living in the soil, freshwater, and the sea, their overwhelming presence becomes obvious. The earliest fossils date back some 460 mya to the Ordovician period, and by the Devonian, terrestrial plants were being invaded by fungi.[273] There are even reports of fossil fungi in ancient *Araucarioxylon* wood, thus providing some credence to our scenario of Cretaceous wood-boring insects transmitting plant-pathogenic fungi in those ancient forests. Fossils of bark and long-horned beetles dating from that time further support this scenario. Other gymnosperms could have lost their leaves from fungal attacks, since epiphyllous types growing on conifer leaves also existed in that period. Moreover, fossils show that by the mid-Cretaceous, mushrooms were quite abundant (color plate 12D),[345] as well as fungal parasites such as one infecting a scale insect in Burmese amber (color plate 13D).

Aside from affecting dinosaurs indirectly by destroying their plant food sources, fungi could have infected and even killed hatchlings, just like they do to reptiles today.[275] Many of these fungi belong to common soil genera (*Aspergillus, Mucor, Cephalosporium, Penicillium, Beauveria,* and *Fusarium*) that either grow on the skin or lodge in the respiratory and digestive systems. One species of *Beauveria* is known to kill alligators,[276] and since this genus also attacks insects, it is possible that arthropods mechanically transmitted the spores. Certainly the most lethal infections to dinosaurs would have been those that became established in their respiratory systems and caused mycotic pneumonia.

### Protozoa

Protozoa, with some 65,000 described species, are an ancient group with a fossil record extending back to the Archean Eon, some 2 billion years ago. While the earliest forms were certainly free-living in the sea, feeding on bacteria, algae, and one another, it didn't take long for them to establish symbiotic associations with other life forms. Their first relationships were probably simple attachments to the body wall of organisms living in the same aquatic environment (ectophoresis). At one point in time, they developed ways to burrow into the skin of their carrier, obtaining nourishment and becoming ectoparasites. Later on, different types invaded the alimentary tracts of animals, just living there and feeding on partially digested food (endophoresis). Many of them, especially the flagellates, became permanently established in the guts of insects, especially in termites where they play an important role in digesting the cellulose (or in carrying around internal organisms that do the job). Additional protozoa, especially the ciliates and coccidians, penetrated into the body cavity of their animal hosts, multiplying mostly by simple division and forming resistant stages (cysts) between hosts. Eventually those that lived in bloodsucking insects were transported into vertebrates, became established, and caused serious diseases. Curiously, very few protozoa, aside from species in the genus *Phytomonas*, formed any continuous associations with plants.

Two important groups were present in the Cretaceous, malarial organisms and trypanosomatids, and fossils show that they have existed for at least 100 million years. Malarial parasites of the family Plasmodiidae consist of ten genera, the best known being *Plasmodium*, which infects reptiles, birds, humans, and other mammals throughout the world.[280–282] While related to coccidian parasites, it is difficult to trace them back to any free-living ancestral group. One Tertiary fossil record of *Plasmodium* plus one Cretaceous record shows that malarial organisms had already evolved their complicated life cycle millions of years ago.[167,283]

The life cycle of these protozoans has become very intricate and involves multiple stages. The vector acquires pre-sexual stages of the parasite from the blood of infected vertebrates. After undergoing sexual maturity in the insect, the malarial organism develops into a cyst that eventually becomes filled with minute thread-like bodies called sporozoites. Upon their release, the motile sporozoites migrate through the arthropod's body cavity and enter the salivary glands, where they are transferred back to the vertebrate at the next blood meal (fig. 23).

The four types of malaria infecting humans all belong to the genus *Plasmodium*, which also infects monkeys and the higher apes. Malaria caused by species of *Haemoproteus*, considered to be the most primitive type, is spread by biting midges and occurs in birds, reptiles, and amphibians. This type of malaria was discovered in Burmese amber.[167] What is difficult to establish is just when biting midges started transmitting these vertebrate pathogens. Since the earliest known biting midge fossils occur in Lebanese amber, and some of these have mouthparts adapted for biting vertebrates, disease transmission could have occurred by that period. Because the effects of these organisms on extant reptiles are poorly known, we can only compare their possible effects on dinosaurs from our knowledge of human malaria, which can be quite devastating.

Flagellates of the family Trypanosomatidae, which were also discovered in Burmese amber, include those responsible for sleeping sickness and leishmaniasis. The motile stages have a flagellum emerging from their anterior end and do not differ significantly in appearance from their supposed free-living ancestors, the euglenoids. The first stage toward parasitism in this group was probably the establishment of colonies in the guts of animals living in the environment. The original hosts may have been marine nematodes, followed by soil and freshwater roundworms. Those trypanosomatids that did eventually colonize the gut of bloodsucking insects such as sand flies could have evolved into the vertebrate-pathogenic leishmanias.[346]

Recent studies indicate that parasitic lines evolved at least four times with these protozoa.[285,286] This is obvious because the fish and amphibian parasitic trypanosomatids vectored by leeches certainly arose independently from those carried by sand flies to mammals and lizards. While the origin of infectious trypanosomatids is controversial, with some suggesting they arose in vertebrates,[284] sand flies possibly obtained the parasites originally from plants infected with *Phytomonas*. The discovery that female sand flies imbibe plant secretions and *Leishmania* produces cellulose-degrading enzymes[287] supports a plant-origin hypothesis for at least one of these lines. All we can be certain of is that sand fly-vectored trypanosomatids infected reptiles 100 mya.

If these vector associations were newly established in the Early Cretaceous, the effect of trypanosomatids on vertebrates could have been devastating. Possibly the dinosaurs were one of the original host groups for these "emerging" pathogens, which often have a severe impact on their hosts.[288] Both *Trypanoplasma* and *Trypanosoma* are transmitted to fish by leeches. Because of a long association with their hosts, species of *Trypanosoma* are no longer lethal to the fish. However, species of *Trypanoplasma*, which are regarded as emergent pathogens that only recently (evolutionary speaking) began infecting fish, can inflict debilitating illness and death. During the early stages of coevolution, sand fly-transmitted trypanosomatids could have behaved like *Trypanoplasma* and caused rampant infections, epidemics, and deaths.

### Nematodes

Of the approximately 20,000 described nematode species, over 6,000 live in vertebrates, about 4,000 are associated with invertebrates, and 3,000 or more are plant parasites.[173,277,278] The remainder are free-living in soil, fresh water, and salt water. At least 39 species parasitize humans, 46 occur in dogs, and 32 in cats. The tissue specificity of nematode parasites is apparent when examining their locations in cats; different ones are found in the colon, stomach, small intestine, kidneys, lungs, heart, spinal cord, lymph vessels, diaphragm muscles, and even the middle ear.[279] Reptiles and amphibians share many more genera of parasitic nematodes than reptiles and birds or reptiles and mammals, suggesting that body temperature is an important factor in host selection.

Nematodes inhabit all possible niches, from deep-sea trenches to the placenta of whales. The earliest described fossils are mermithid parasites of insects dating from the Early Cretaceous (color plate 13C), but other fossils extend back to the Devonian. Ancient marine roundworms probably swarmed through the oozes bordering Cambrian seas since those inhabiting marine and freshwater habitats today represent the most primitive lineages. Nematodes presumably adapted to terrestrial habitats by the Silurian, some 420 million years ago. These free-living forms could have easily shifted their diets from marine to freshwater and soil microbes, ultimately invading available niches in plants, invertebrates, and vertebrates. Curiously, many did not give up their early dependence on bacteria, and the first animal parasites were probably pinworms (Oxyurida) that feed on bacteria in the alimentary tract of invertebrates and vertebrates. Other parasites (Strongylida) switched to a fluid diet in the vertebrate alimentary tract, but the free-living juveniles still feed on soil bacteria before they enter the host. Some nematodes, such as those in the insect-parasitic genera *Steinernema* and *Heterorhabditis*, solved their dependency on bacteria by introducing specific species of soil bacteria into their hosts. The bacteria multiply and kill the insects, at the same time providing the worms with a food source inside a protected microenvironment.[279]

Nematodes continued to coevolve with their plant, invertebrate, and vertebrate hosts, probably establishing parasitic relationships soon after their hosts appeared. The stomach worms (ascarids) were extremely successful in severing their dependency on bacteria. Aside from infecting over a billion humans as well as other mammals, ascarids also parasitize sharks, fish, amphibians, reptiles, and birds[279] (fig. 32). They even parasitized dinosaurs[135] (color plates 16C, 16D).

Two separate lines of vertebrate parasitic nematodes, the spirurids and filarids, use arthropod vectors to reach their hosts.[173,279] Filarids develop in the tissues of their hosts and produce special micro-larvae (microfilaria) (fig. 31) that are picked up when the vectors take a blood meal. In contrast, spirurids live in the alimentary tract of vertebrates. Their eggs are voided with the feces and hatch when eaten by various beetles and orthopterans. The roundworm's future depends on a vertebrate choosing the parasitized insect for lunch. A large number of vertebrate parasitic nematodes occur throughout the world today, and some of those whose ancestors could have parasitized dinosaurs are discussed in the chapter on parasitic worms.

We know essentially nothing about plant-parasitic nematodes of the Cretaceous world. But be assured that they were just as common as they are at present and attacked the roots and in many cases the stems and leaves of representatives of all the diverse floral groups that thrived during that period.

So just how badly did dinosaurs suffer from parasites? Based on what we know regarding pathogens of reptiles and birds and when the various groups appeared in the past, dinosaurs were susceptible to oral sores caused by bacteria, trematodes, and nematodes, lesions in their esophagi from kalicephalid nematodes, stomach wounds from spirurid and ascarid nematodes, and intestinal damage from capillarid nematodes, *Ophiotaenia* tapeworms, and amoeboid protozoans. Their internal tissues probably contained trematodes and filarial nematodes, and their muscles tapeworm larvae (plerocercoids) and nematodes. The blood vessels, especially those around the heart, in all likelihood were filled with filarial nematodes. Microfilaria undoubtedly circulated through their veins, and malarial and leishmanial parasites developed in their blood cells. Perhaps their lungs were filled with protozoans and lungworms and their spleen and liver contained malarial and leishmanial pathogens. Even their fat tissue conceivably hosted trematode larvae and protozoans. In addition, fly larvae could have developed in their nostrils as well as around scratches or wounds. And lastly, a plethora of viral and fungal diseases certainly were responsible for everything from pneumonia and diarrhea to death. This would have been the worst scenario.

Diseases can be very important controlling factors of animal populations, although they are often overlooked.[66] Parasites affect their host in many ways, from behavioral and physical changes to a reduction of fecundity, and finally death. There is ample evidence that vector-borne diseases can inflict high mortality levels, especially if they recently, geologically speaking, appeared on the scene like some human forms of malaria. If sand flies first established vector relationships with trypanosomatids, and biting midges with haemoproteans, in the Early Cretaceous, then these new diseases definitely were capable of threatening the very existence of dinosaurs. Even if some were able to tolerate full-blown infections, other stress factors such as starvation or additional parasites would have upset their equilibrium and led to death.

Global epidemics potentially occurred as infected sand flies, biting midges, and dinosaurs dispersed and expanded their range. Pandemics possibly spread between Asia and North America via the Beringia land bridge, which was present by the Late Cretaceous.[289] Since arthropod carriers were undoubtedly universally distributed at that time, the disease cycle continued no matter where dinosaurs lived. Further distribution of diseases between and in North and South America became possible through Central America in the latest Cretaceous.[290] Infected insects then as now were dispersed by wind currents, and mosquitoes and midges have been collected at altitudes from 5,000 to 13,000 feet.[291] At such heights, intercontinental dispersal of pathogen-bearing insects becomes a definite possibility.

Just as we now fear flu viruses that can be introduced by migrating birds and then "jump" to humans, the same scenarios were probably happening in the Cretaceous with other diseases.

# 22.

## Insects: The Ultimate Survivors

INSECTS HAVE been around for more than 400 million years. Dinosaurs (non-avian) only lasted 180 million. What determines how long families, genera, and species survive? When biological and physical events impact a species so that the death rate continuously exceeds the birth rate, that life form has begun a downward spiral towards extinction. Ultimately a low population threshold is reached where recovery is impossible and the fate of the species is sealed. That loss has a ripple effect throughout the ecosystem, the severity of which depends on the importance of the species. Demise of a keystone species invokes the greatest aftermath; the cessation of one with global distribution has worldwide consequences, producing a cascade effect throughout the species range. Depending on the organism's importance, the effect can be short or long term, but nothing disappears from the earth without producing some change.

To see how the cascade effect works, let's assume that a species of large tree in a tropical rain forest vanishes in a short period of time, say 100 years, due to a fungal infection. It was a keystone species, a dominant canopy tree that supported a wide range of flora and fauna. The first to follow are those specialists dependent on that tree's roots, foliage, flowers, fruits, seeds, and so on. These would probably be small life forms such as insects and fungi. Some trees host 1,200 or more invertebrates, and one-tenth of them could disappear. That is an immediate effect. But there are more gradual effects. The loss of the species to the ecosystem causes competition and changes in the community composition. Other plant species may die, taking their associated flora and fauna with them. Animals that fed on the invertebrates might move on to other prey, upsetting the balance and forcing more changes to occur. Event A leads to events B, C, D, and these in turn lead to E, F, G, H, I, and so forth. The overall magnitude of the events can never de determined. In the normal course of life, species composition in communities oscillates, a factor that leads to evolution and diversity. Sometimes, the disappearance of a species is not a terminal event but a pseudoextinction, which occurs when a plant or animal undergoes a gradual phyletic transformation into a new one. The organism never really disappears, but slowly becomes a daughter species. This gradual transformation provides its associates time to adapt and has minimal consequences to the ecosystem.

Extinctions are generally classified as background or mass. Background extinctions vary in intensity over time and are stochastic—the result of normal, random population changes. The rate of species loss from background extinctions has been estimated to vary from a hundred to thousands per year. Mass extinctions occur when a large percentage of the global population of plant and animal species (usually greater than 50%) die off abruptly. These tend to be more deterministic in nature, a fate that is inescapable though adaptation. All species of plants and animals today are the result of a gradual change of ancestral forms into modern descendants. Species found as fossils in the Cretaceous have all disappeared as the result of background, mass, and pseudoextinctions.

Natural selection infers that some determinants exist in species that makes them susceptible or resistant to background or mass extinction when confronted with a particular combination of biotic and/or abiotic circumstances. Some are listed in appendix B. Examples of biotic (biological) factors are the ability or inability to adapt, ward off infections, disperse, or feed on a wide range of foods. Abiotic (physical) factors are usually environmental, such as a rise or drop in temperature or loss of habitat. An animal or plant may be viewed as vulnerable to extinction when faced with more negative factors than positive ones. The question becomes, how many adversities can a species take before it succumbs?

Insects have been remarkable survivors over geological time, and their success can be attributed to a cumulative evolutionary selection of characters that shield them from extinction. To what do they owe their good fortune? They have been able to prosper since their origin in the Paleozoic because they had the advantage of adapting certain survival strategies, such as small size and vast diversity. Compare this with non-avian dinosaurs that elected for large size and limited diversity. The combination of just these two factors can affect the overall survival of genera and families. Because of their extreme diversity, insects can afford to have many more specialists. And small size makes it easier to find adequate food and shelter.

Another feature we should stress, besides their immense biomass, is their amazing range of reproductive strategies, including the production of large numbers of eggs and short generation times. Insects can complete one (univoltine), two (bivoltine), or multiple (multivoltine) generations every year. In general, smaller insects have more generations than larger ones. The most common form of reproduction is oviparity, where fertilized eggs are deposited in the environment. However, there is also ovoviparity (eggs hatch within the female and active larvae are deposited), viviparity (living young are produced), paedogenesis (reproduction by larval forms), polyembryony (one egg divides to produce multiple eggs), and finally hermaphroditism and parthenogenesis in the absence of males.

Under ideal conditions some flies can develop and reproduce rapidly, with 25 or more generations per year. It has been estimated that a pair of *Drosophila* flies could produce $1.19 \times 10^{41}$ offspring in a single year.[292] A housefly pair could generate $1.91 \times 10^{20}$ progeny in a single summer, and honeybee queens are thought to lay 1,500 to 2,000 eggs in a single day.[293] Although the number of eggs varies from a few to thousands, the majority of insects deposit between fifty to several hundred at a time.

These reproductive strategies differ markedly from those suggested for dinosaurs. The majority, if not all, dinosaurs laid eggs. Unfortunately there is no evidence to suggest how frequently these events were. Some are thought to have been viviparous.[294] Inference from birds and reptiles suggests they had only one brood per season, and fossils indicate that the number of eggs varied from only 2 or 3 in many species to 20 in some hadrosaurs and even 26 in the case of one tyrannosaur. The number produced over a lifetime is unknown and probably varied with life span. The period from egg to reproductive adult is also a mystery, although estimates of 8–10 years in hadrosaurs and up to 62 years in sauropods have been suggested.[295] The shorter the generation time and the more offspring produced each lifetime, the better able a species is to adapt to unfavorable environmental conditions through mutation and natural selection.

The ratio between egg or hatching weight and adult weight, an indication of how much organisms have to grow to reach reproductive age, can be an important factor in survival. This ratio varies among insects because the eggs of exopterygotes (those with external wing development) are larger that those of endopterygotes (insects with internal wing development).

Dinosaur egg weights were a small percentage of adult weights, especially with large dinosaurs. Eggs of hadrosaurs have been estimated to weigh about 2 pounds, and while some adults reached 4,400 to 8,800 pounds, the largest weighed in at 17.5 tons. That means an adult could have been 2,200 to 17,000 times bigger than its hatchling. Putative monumental dinosaur eggs (probably from the sauropod *Hypselosaurus*) have been described as 12 inches long and 10 inches wide.[114] This is similar to those of an extinct ostrich (Aepyornithidae), which had eggs 3 feet in circumference, or nearly a foot in diameter, with an estimated liquid content of two gallons.[335] Even with eggs of such immense size, the adult would still be more massive than the hatchling by quite a bit.

Insects frequently lay their eggs in sheltered areas and/or enclose them in a protective material provided by the mother, such as cases or nests of body hairs. Some are laid in cracks and crevices in the ground or inside plant or animal hosts. Many are attached to a food source. Most hatch as precocial young, which are on their own after birth. With rare exceptions, insects do not provide any care to their young.

There is evidence that at least partial parental care existed with some dinosaurs. Fossil eggs have been found in nests that were buried and others in depressions made in soil and vegetation. This and other data suggests that nesting activities may have ranged from actual brooding to protection and feeding of the young in colonies. However, the debate over whether the youngsters were precocial and hatched to fend for themselves or altricial and required parental care is still ongoing, and a range of behaviors is likely. If dinosaur hatchlings suffered attrition rates seen in many extant ground-nesting animals (90% or more the first year), and had to attain an adult size some five hundred to several thousand times their birth weight with little or no parental care before reproducing, then the survivors represented quite an achievement.

One cause suggested as contributing to dinosaur extinction is a temperature- sensitive sex determination factor such as occurs in crocodiles and alligators.[296] When crocodile eggs are incubated at temperatures below 31.7°C, they produce females; from 31.7°C to 34.5°C the result is males, and above 34.5°C only females appear. There is no evidence that such a mechanism existed in any dinosaur, but if so, it may have become a significant aspect in endangered populations. Most insects, on the other hand, are not known to have temperature-dependent sex-determinate factors.

Insects have many additional survival strategies that have contributed to their overall success. One of the most important is their ability to enter diapause, a prolonged dormancy that is hormonally regulated. Diapause can either be obligatory or facultative. The first type occurs irrespective of environmental stimuli, while the second is tied to them. In the tropics, these stimuli may be drought, high temperatures, or food shortages, while in more southern or northern latitudes a shorter photoperiod triggers the onset of diapause. No mechanism similar to diapause or even aestivation or hibernation has been suggested for dinosaurs.

Diapause affords insects the ability to survive environmentally stressful times. The condition is associated with a decrease in oxygen consumption and a lowered metabolism. Oxygen is normally carried to insect tissues by diffusion via a system of cuticle-lined tubes and branches called tracheas and tracheoles. Smaller insects may lack trachea and simply depend on diffusion across the cuticle. While this type of breathing bypasses the use of blood hemeproteins, it does limit the insect's size. The capacity to obtain oxygen by diffusion and regulate oxygen consumption contributes to their survival.

Breathing mechanisms for large dinosaurs are still a matter of speculation because no fossilized lung tissue has been found. Debate on whether their lungs were bird-like or crocodile-like continues. Whatever the mechanism for getting air to the tissues, the organs must have been immense to accommodate the oxygen requirements of a 50- or 100-ton sauropod. On the other hand, the needs of a 400-pound dinosaur would be considerable lower. Also, ectothermic animals that take on the temperature of the surroundings and endothermic animals that maintain their own body temperature would have different oxygen requirements. Since the issue regarding thermoregulation in dinosaurs is still being contested, lung size and function cannot be resolved.

A significant feature of insects that makes up for small size, aids in their dispersal, and gives them the ability to exploit many niches is the power of flight. If you have ever tried to catch a fly, you know about their extreme maneuverability and speed, as well as their ability to move sideways and backward in flight. About 90% of insects fly in the adult stage.

What were the aeronautical abilities of dinosaurs? Those who feel birds evolved from theropods could use the few Cretaceous birds as examples of a dinosaur line exploiting the air with insects. One theory for the origin of bird flight suggests it was the result of ground-dwelling pre-avian dinosaurs leaping into the air in pursuit of flying insects. Today about 80% of warm-blooded vertebrates fly (900 species of birds and 1,000 of bats).[297] However, the great majority of Cretaceous dinosaurs probably did not have the power of flight.

An overall comparison between dinosaurs and insects illustrates differences between what are called $r$ and $K$ strategists. The $r$-species are small organisms with rapid growth, high reproductive rates, and short lives. The $K$-species are those with slow development, large body size, low reproductive rates, and long lives. Mortality of the former is usually independent of population density, while in the latter, it is frequently density related. Also, the former characteristically occur in variable or unpredictable environments, while the latter thrive under uniform or predictable conditions. Looking at the animal world today, it is obvious that $r$-selected populations are more diverse and successful than $K$-selected populations.[66] Large dinosaurs represent the ultimate $K$-strategists while insects are the epitome of $r$-strategists (appendix B).

The list of adaptive features in insects that contribute to their success is extensive. In fact, there are few things about them that don't spell success. But one of the most amazing accomplishments of some is their ability to feed above their trophic level. In the normal scheme of things, animals only feed on others smaller than themselves unless they hunt in packs. Thus wolves can kill a 500-pound caribou. Insects, however, have jumped the scale and feed fearlessly on animals thousands of time their size—herbivores, carnivores, it doesn't make any difference to them, and this makes them the top predator in any ecosystem! They don't usually kill: there are only a few records of animals dying from exsanguinations by mosquitoes and blackflies. However, they can indirectly cause their host's death from what they carry, disease-causing pathogens.

# 23.

## Extinctions and the K/T Boundary

EXTINCTIONS vary in intensity from normal low-level background to major events where some 50% of the total species of plants and animals disappear. Of the many global restructurings that have been detected, only five qualify as mass extinctions. The best publicized occurred around the Cretaceous-Tertiary (K/T) boundary some 65.5 mya and coincides with the demise of the dinosaurs. The many theories that have been proposed for the cause of these extinctions have broadly divided the scientific community into two camps, the catastrophists and the gradualists, and pivot around the central issue of time span. How fast did the detected extinctions take place? Was the time frame overnight, over centuries, millennia, or millions of years? Unfortunately, the fossil record is fraught with problems and inconsistencies that make it difficult, if not impossible, to interpret the timing and magnitude of extinctions. Some of the more obvious difficulties are listed in appendix C.

What do the catastrophists say? They tend to favor a mass extinction that was instantaneous or occurred over an unspecified but short time, geologically-speaking. Their theory involves the meteor (or comet) impact proposed by Alvarez and his group.[336] Ample physical evidence has been collected to indicate that an impact occurred in the sea,[9] including an iridium spike in the soil at the K/T boundary, shocked quartz, tektites, and a crater of the correct size at Chicxulub in the Yucatan Peninsula.

The gradualists do not deny that an impact occurred, but they also consider other physical events that occurred around the K/T boundary, including marine regression—a drastic decline in sea level—increased volcanism, and climatic changes. Many feel that it's either too early to say or that the available facts are insufficient to support either an abrupt and/or mass extinction. The gradualist's explanations tend to favor a series of accelerated background extinctions—a random grouping of coincidental events before and at the K/T boundary that had a cumulative, global effect.

Perhaps the best way to illustrate the necessity of reappraising the terminal Cretaceous extinctions would be to examine today's world. Keep in mind that by definition, at least 50% of all species must have perished for a mass extinction to occur. If an extinction event happened right now and killed off *all* terrestrial and aquatic animals *except* insects, that loss would only be about 25% of the total animals. If 50% of the vertebrates died out, as was suggested to happen at the K/T boundary, it would represent only about 2% of *all* animals. To a vertebrate zoologist, the loss would be significant, and the rest of us would probably concur because after all, we are vertebrates, but that would not make it a mass extinction.

How abrupt was the demise of those organisms that appeared to die out at the K/T boundary? The answer depends on the particular scientist's viewpoint. The literature abounds with contradictions centering on who, where, and when. Many of the groups whose populations were examined around the K/T boundary were marine forms, and it has been estimated that approximately 38% of the animal genera in the sea became extinct at the K/T boundary,[298] even though most of the major families survived. Genera that disappeared included some marine reptiles, certain lines of bony fish, sponges, snails, clams, cephalopods, sea urchins, and foraminifera, particularly those from shallow tropical seas. However, many of these appear to have already been declining in diversity prior to the K/T boundary. Among the giant marine reptiles, ichthyosaurs disappeared well before the event, plesiosaurs were decreasing at the time, but the mosasaurs appeared to have been thriving 65.5 mya.[299]

Many of the marine groups that did disappear, such as the

ammonites and the belemnites, both cephalopods, and the microscopic foraminifera, were also declining far before the K/T boundary, indicating that other factors threatened their existence. Thus these extinctions cannot be directly tied to a catastrophic impact at the end of the Cretaceous. If some of these threatened groups were keystone species, then their demise was undoubtedly directly linked to the extinction of other organisms that depended on them for survival. Such a scenario possibly occurred with the demise of rudist clams, the major component in Cretaceous reef architecture.[300] The loss of reef habitats due to their decline and disappearance would have eliminated an enormous number of codependent species that depended on such environs for survival. The cause of rudist clam extinctions may have been as simple as a disease intensified by marine regression and accompanied by changes in ocean currents and temperatures.

Similar inconsistencies have been noted with terrestrial plants, the very foundation of the food chain. The data regarding K/T plant extinctions is conflicting but suggests a global pattern. Many scientists agree that there was a latitudinal effect because, based on spore and pollen remains, vegetation in the more southern latitudes (Australia, New Zealand, and Antarctica) was relatively unaltered. The more northern latitudes also indicated fewer floral turnovers than mid-latitudes, and even regional differences were noted. Much of the data reveals variable species turnover before and at the K/T boundary. And it appears that angiosperms were the most affected by events of the K/T transition. The most severe effects on plants have been reported from the Hell Creek Formation in the western interior of the United States. Changes in fossil leaf morphology there were interpreted as evidence of a 79% turnover in plants at the K/T boundary, but this was *after* a 75% turnover observed prior to that. So plants in that area were already in a state of flux. In contrast, palynoflora (pollen) data from the same area imply only a 30% plant extinction rate.[301,302] Another pollen study also indicated that species were changing before the K/T boundary, as plant pollen became unusually small. This was interpreted as evidence that ecosystems were transitioning and that plants were probably at the edge of their preferred geographic range as well as stressed from climatic changes, as the community structure went from open to closed habitats.[303] In contrast, in the Raton Formation of New Mexico and Colorado, 85% of the plant species are thought to have survived,[304] which is similar to the 70% proposed survival rate in Saskatchewan.[305]

This apparent variability in plant extinctions may be attributable to there being very few fossil sites with plants and dinosaurs that span the Cretaceous-Tertiary boundary. Presently these are confined to the western interior of North America, although sites in South America and Asia hold promise for contributing additional information. Those at the Hell Creek/Lower Tullock formations in eastern Montana have been used extensively to gauge the severity of extinctions at the K/T boundary. One problem with this limited data pool is that the changes noted there might be an anomaly. Since extinctions seen in leaf fossils around the K/T boundary at Hell Creek do not seem to be supported by information gleaned from studies in more southern and northern global latitudes, nor by the pollen record, one is left to wonder if the changes there were related to some short-term localized phenomena.

One of the abiotic causes for regional extinctions could have been drought. Some data has suggested that the Hell Creek area was seasonally much drier than originally thought. To see how catastrophic a drought can be, consider the most well-known one in the United States. It created the infamous Dust Bowl that devastated the Great Plains in the 1930s, lasted more than a decade, and by 1934 had affected 75% of the country and twenty-seven states.[306–308] Lack of water was exacerbated by overgrazing and by replacing deep-rooted native plants with shallow-rooted agricultural crops. Hot, dry windstorms picked up the topsoil and created dust clouds so thick that it was dark at noon. The windborne soil was carried over 1,500 miles away. About 90 million acres (140,625 mi$^2$ = 3.64 × 10$^5$ km$^2$) of farmland was destroyed or severely damaged. How amazing that this amount of habitat destruction occurred in such a geologically insignificant time frame. Facts have been presented indicating that the drought was caused by anomalous tropical sea-surface temperatures and amplified by interactions between the atmosphere and the land surface.[307] So it is easy to image that droughts, dinosaur over-grazing, and a dearth of deep-rooted plants might account for the successive turnovers in fossil vegetation noted at Hell Creek. Other scenarios could also explain what are perceived as drastic localized extinction events, such as periodic incursions and re-gressions of the inland seas that destroyed old habitats and cre-ated new ones. These possibilities underscore the necessity for investigating many other vertebrate and plant fossil sites that span the K/T boundary globally before arriving at definite con-clusions about extinctions.

Estimating insect extinctions at the K/T boundary is hindered by gaps in the fossil record throughout the Late Cretaceous and Early Tertiary. There are currently no sites with enough fossils bridging the K/T boundary to determine if and when various groups disappeared. Second, insects were so diverse that to ob-serve a 50% loss of species would be impossible with the avail-able fossils. Knowing that these caveats exist, insects appear to have passed the K/T boundary relatively unscathed.[309–312] This is based on the examination of higher taxa, especially families, that persisted into the Tertiary. Since an insect extinction event was observed in the mid-Cretaceous using these taxa levels,[35] if one of a similar magnitude occurred at the K/T boundary, it should be detectable. Since insect species represent the majority of ani-mal diversity, why haven't they even been mentioned in most discussions of mass extinctions?

Ninety-six percent of all animal life are invertebrates,[309] and this figure does not even include unicellular or microbial organ-isms like protozoa. A large portion of those are soft-bodied and not amenable to fossilization, so any population changes at the K/T boundary cannot be assessed. Others with a reasonable Ter-tiary fossil record, such as spiders, are poorly represented in Cretaceous deposits. Those with a poor fossil record, such as the mites and the lesser-known fossil arachnid groups (scorpions, daddy longlegs, wind spiders, whip scorpions, pseudoscorpions), as well as millipedes and centipedes, also have not been considered in estimating losses at the K/T boundary. Thus, aside from the dinosaurs, only a few marine invertebrate groups with hardened parts have been used to evaluate K/T extinctions. Perhaps at this point in time, with the incompleteness and bias of the fossil record, scientists need to reevaluate the data. Since the fossil record of common invertebrate groups is so poorly known, estimating possible extinctions at the K/T boundary has to be based on their present diversity and distribution.[320] An example is the nematodes, extremely important organisms in the ecosystem not only as animal parasites but because they are responsible for 50% of the energy flow in the soil, and their feeding can decrease above-ground plant production by 30–60%.[66]

In numbers of individuals, nematodes are probably one of the most abundant invertebrate groups on the face of the earth, ranging from 3 billion in an acre of soil and 1.5 billion in the upper 20 mm of an acre of marine beach sand[313] to 9 million in a square meter of temperate grassland.[314] As many as 71 species have been recovered from just 100 grams of grassland soil.[315] While only some 20,000 species have been described, estimates of their diversity range from 500,000 to 10 million species.[279,316] They occur in soil, fresh water, and the marine environment as well as inside plants and animals. Because of the cosmopolitan distribution of so many genera today, it is obvious that many terrestrial microbotrophic (utilizing microbes as food) as well as plant and animal parasitic nematode lineages[317–319] extend back to the period when all continents were united into the supercontinent Pangaea. Yet their fossil record is virtually nonexistant.

There is no doubt that large non-avian dinosaurs died out sometime around the K/T boundary. Debate still rages whether they were already experiencing a loss in diversity by the Late Cretaceous. Some consider that both genera and species declined dramatically in the Maastrichtian by about 40%,[321,343] while others argue for no loss and perhaps even an increase in diversity,[322,323,342] and some suggest that dinosaurs were holding their own.[324] While the dictum is that all dinosaurs died out at the K/T boundary, there are several reports of dinosaur fossils after the Cretaceous. One well-documented study provides convincing evidence that some fossils date from the Paleocene and were not redeposited. Bones of ornithomimids, dromaeosaurs, saurornithoidids, ankylosaurs, nodosaurs and hadrosaurs were recovered in the Ojo Alamo beds of the San Juan Basin in New Mexico, not more than 965 miles from the Chicxulub meteor impact site.[325] But how could large dinosaurs survive so close to ground zero? The authors suggest that they endured as buried eggs that hatched after the impact event. But if dinosaurs did survive into the Paleocene, how much longer did they last?

All of the proposed events impacting the planet's ecosystems over a ten-million-year time span around the K/T boundary would have had similar drastic consequences. They resulted in physical changes in the land surface. The marine regression that occurred from 70–60 mya culminated in an 11.2 million $mi^2$ loss of ocean primarily in shallow neritic areas and epicontinental seas, with a concomitant increase in land.[9] The Deccan Traps volcanism originally resulted in an approximately 772,200 $mi^2$ basaltic flow up to 7,872 ft deep across parts of India that accumulated 69–63 mya.[339] And the meteor impact on the Yucatan peninsula right at 65.5 mya carved an underwater crater with an area estimated to range from as little as 2,826 $mi^2$ to as much as 27,300 $mi^2$ and threw debris into the atmosphere.[340,341] The volcanic activity and the meteor impact caused some similar consequences: release of heat, air pollution with increases in $CO_2$ and $SO_2$, toxic fallout including acid rain, ozone damage, diminished sunlight, and contaminated water. The difference is that the Deccan Traps volcanism was episodic over 6 million years, while the meteor produced an environmental impact that lasted for a much shorter period.

The three major events surrounding the K/T boundary produced some common repercussions, including changes in ocean temperatures and salinity, as well as in global wind patterns and worldwide terrestrial temperatures. In addition, marine regression was accompanied by changes in ocean currents, wave action, and the addition of 11.2 million mi$^2$ of land.[9] The latter brought about shifts in coastlines, the formation of land bridges, and an increase in freshwater lotic systems as the distance that rivers and streams flowed to reach the seas extended.

Some areas may have been more impacted than others, while effects such as climate change would have been worldwide (fig. 39). Many of the results would have been cumulative, with all working in tandem to produce a greater outcome than one event. Obviously the planet was suffering from a series of environmental stresses for an extended period of time, and these would have affected all life forms.

Although most generally accepted theories for the causes of the Cretaceous-Tertiary extinction events center around marine regression, meteor impact, or volcanism, many other conditions[326] could have added to the stress load of endangered species, thereby decreasing their vigor and making them susceptible to extinction.

Abiotic factors would have had diverse and harmful effects on all life forms. Just as climatic change can be considered a direct abiotic corollary to these three events, disease is certainly a biotic corollary shared by them. Anytime an ecosystem is challenged by harmful physical events, the organisms within it experience stress, virulent pathogens appear, and epidemics run rampant.

We believe that disease played a significant role in dinosaur extinction during the terminal Cretaceous. The importance of diseases in shaping ecosystems should not be overlooked. Animals can be brought to the brink of extinction by disease, and a combination of other factors can institute, contribute to, or finish off the process.

We are not talking about a single disease wiping out all dinosaurs, but rather the cumulative, cascading effects of many diseases working along with major contributing abiotic elements and diverse biotic factors such as those discussed in previous

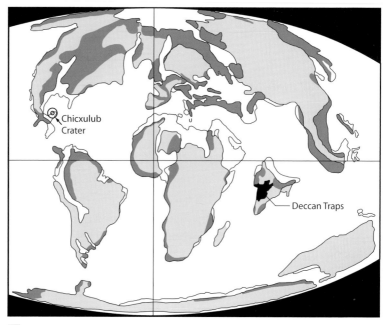

■ Areas of marine regression
░ Land

FIGURE 39. This map illustrates the probable sites of the confluent events 60–70 millions years ago (mya) that framed the K/T boundary. The configuration of the continents and their positions is based on a comparison of the Maastrichtian and Paleocene maps of Smith et al.[9] Light gray represents exposed, old land. Dark gray represents newly acquired land due to marine regression from 70 to 60 mya. These same darkened areas then also represent loss of neritic (continental shelf) seas, the primary source of life and energy production in the oceans. The extant of the basaltic floods associated with the Deccan Traps is shown in black.[339] The putative site of the meteor impact on the Yucatan Peninsula, the Chicxulub crater, is ringed. This site was apparently submerged from the Aptian (120 mya) to the Late Eocene (37 mya) and therefore at the time of impact. The present-day outlines of the continents are shown. Extensive areas in Africa, Europe, northern India, central North America, and north-central South America were submerged.

chapters. Others have suggested that diseases could have contributed to Late Cretaceous extinctions. Over ten years ago, Robert Desowitz, a medical parasitologist, suggested that an epidemic of reptilian kala azar (an especially lethal form of leishmaniasis) transmitted by sand flies could have caused the extinction of the dinosaurs.[327] But there was no evidence of any Cretaceous pathogens at that time. However, amber studies have shown that vertebrate pathogens were indeed present in the distant past and could have been significant causes of mortality.

As we have pointed out, the effects of *Leishmania* and malaria on dinosaurs could have been quite severe, especially if they already had compromised immune systems. A current example of how pathogens act in concert is *Leishmania* and HIV coinfections in humans. The synergistic effect of these two maladies causes a very serious new disease complex since the HIV virus weakens the patients' immune system and allows the *Leishmania* to spread unimpeded into internal tissues.[328]

In today's world, pathogens have been implicated in the demise of many species, and such parasite-driven host extinctions can occur in populations of any size, especially when host reproduction is significantly reduced.[329] The decline and extinction of frogs in tropical American and Australian rain forests are now considered the result of fungal diseases enhanced by a slight rise in global temperatures.[330,331] The devastating bluetongue disease of ruminants has spread into northern Europe within the past decade as a result of climatic changes, allowing the virus to be acquired and transmitted by endemic species of biting midges and resulting in the deaths of over one million sheep.[165]

The recent extinctions of endemic Hawaiian birds by malaria show how damaging a disease can be when introduced, along with its vector, into populations of immunologically naïve hosts.[332] A land snail in the Pacific Islands was driven to extinction by a microsporidian parasite.[333] Examples of introduced pathogens causing the decimation of human populations include measles and mumps in Native Americans, as well as pandemics of plague, typhus, Asian flu, and yellow fever. Pathogens are constantly mutating and producing new strains capable of infecting different host groups. That is why a global pandemic of the avian flu virus is considered likely in wild mammals, birds, and even humans if an especially pathogenic mutant appears.[334]

The world was changing as the Cretaceous came to an end. The spread of flowering plants pollinated by insects modified the ecosystem, insects competed with dinosaurs for food, and insects and other arthropods transmitted emerging pathogens. At the same time there were periods of temperature change, marine regression, volcanic eruptions, and one or more meteor impacts. Such a confluence of biotic and abiotic events would have been a perfect setting for the spread of diseases.

Although it is difficult to believe that microscopic organisms could render one of the final knockout blows to such powerful giants as the dinosaurs, it nevertheless stands as a feasible scenario when viewed in the overall context of already threatened and debilitated populations. Whether a catastrophist or gradualist, you cannot discount the probability that diseases, especially those vectored by miniscule insects, played an important role in exterminating the dinosaurs.

# Appendix A

## Cretaceous Hexapoda

RECORDS from various sources, especially *History of Insects*,[35] provide a total of 32 orders and 490 families. A dotted line means the presence of a group during all or part of the Cretaceous. The letters L, B, and C refer to records from Lebanese, Burmese, and Canadian amber. Asterisks indicate extinct groups. Since new Cretaceous insect fossils are being described continuously, this list is not complete.

| Order | Family | Lebanese | Burmese | Canadian |
|-------|--------|----------|---------|----------|
| | | 145.5 mya | | 65.5 mya |
| **Collembola** | Entomobryidae | - - - - - - - - - - - - - - - - - - - - - - - - - - - - - - - |
| (springtails) | Hypogastruridae | - - - - - - - - - - - - - - - - - - - - - - - - - C - - - |
| | Neanuridae | - - - - - - - - - - - - - - - - - - - - - - - C - - - |
| | Poduridae | - - - - - - - - - - - - - - - - - - - - - - C - - - |
| | Protentomobryidae* | - - - - - - - - - - - C - - - |
| | Sminthuridae | - - L - - - - - - - - B - - - - - - - - C - - - |
| **Machilida** | Machilidae | - - - - - - - - - - - - B - - - - - - - - - - - - |
| (bristletails) | Meinertellidae | - - L - - - - - - - - - - - - - - - - - - - - - |
| **Lepismatida** | Lepidotrichidae | - - - - - - - - - - - - - - - - - - - - - - - - - |
| (silverfish) | Lepismatidae | - - - - - - - - - - - B - - - - - - - - - - - |
| **Odonata** | Aeschnidiidae* | - - - - - - - - - - - - - - - - - - - - - - - - - |
| (dragonflies) | Aeshnidae | - - - - - - - - - - - - - - - - - - - - - - - - - |
| | Campterophlebiidae* | - - - - - - - - - - - - - - - - - - - - - - - - - |
| | Cretacoenagrionidae* | - - - - - - - - - - - - - - - - - - - - - - - - - |
| | Gomphidae | - - - - - - - - - - - - - - - - - - - - - - - - - |

| Order | Family | Lebanese | Burmese | Canadian |
|-------|--------|----------|---------|----------|

|  |  | 145.5 mya |  | 65.5 mya |
|-------|--------|-----------|---------|----------|
|  | Hemeroscopidae* | - - - - - - - - - - - - - - - - - - - - - - - - - |  |  |
|  | Hemiphlebiidae* | - - - - - - - - - - - - - - - - - - - - - - - - - |  |  |
|  | Karatawiidae* | - - - - - - - - - - - - - - - - - - - - - - - - - |  |  |
|  | Petaluridae | - - - - - - - - - - - - - - - - - - - - - - - - - |  |  |
|  | Protomyrmeleontidae* | - - - - - - - - - - - - - - - - - - - - - - - - - |  |  |
|  | Protoneuridae* | - - - - - - - - - - - - - - - - - - - - - - - - - |  |  |
|  | Sonidae* | - - - - - - - - - - - - - - - - - - - - - - - - - |  |  |
|  | Tarsophlebiidae* | - - - - - - - - - - - - - - - - - - - - - - - - - |  |  |
|  | Thaumatoneuridae* | - - - - - - - - - - - - - - - - - - - - - - - - - |  |  |
|  | undetermined | - - L - - - - - - - - - - - - - - - - - - - - |  |  |
| **Ephemeroptera** | Ametropodidae |  |  | - - - - - - - |
| (mayflies) | Australiphemeridae |  |  | - - - - - - - |
|  | Baetidae | - - L - - - - - - - - - - - - - - - - - - - - |  |  |
|  | Behningiidae* | - - - - - - - - - - - - - - - - - - - - - - - - - |  |  |
|  | Epeoromimidae* | - - - - - - - - - - - - - - - - - - - - - - - - - |  |  |
|  | Ephemerellidae | - - - - - - - - - - - - - - - - - - - - - - - - - |  |  |
|  | Euthyplociidae* | - - - - - - - - - - - - - - - - - - - - - - - - - |  |  |
|  | Heptageniidae |  |  | - - - - - - - |
|  | Hexagenitidae* | - - - - - - - - - - - - - - - - - - - - - - - - - |  |  |
|  | Leptophlebiidae | - - L - - - - - - - - - - - - - - - - - - - - |  |  |
|  | Mesonetidae* | - - - - - - - - - - - - - - - - - - - - - - - - - |  |  |
|  | Oligoneuriidae | - - - - - - - - - - - - - - - - - - - - - - - - - |  |  |
|  | Palaeoanthidae* |  |  | - - - - - - - |
|  | Palingeniidae | - - - - - - - - - - - - - - - - - - - - - - - - - |  |  |
|  | Polymitarcyidae |  |  | - - - - - - - |
|  | Potamantidae* | - - - - - - - - - - - - - - - - - - - - - - - - - |  |  |
|  | Prosopistomatidae* | - - - - - - - - - - - B - - - - - - - - - - - - |  |  |
|  | Siphlonuridae | - - - - - - - - - - - - - - - - - - - - - - - - - |  |  |
|  | Torephemeridae* | - - - - - - - - - - - - - - - - - - - - - - - - - |  |  |
|  | undetermined | - - - - - - - - - - - - - - - - - - - - C - - - |  |  |
| **Blattodea** | Blattellidae | - - L - - - - - - - B - - - - - - - - - - - - |  |  |
| (cockroaches) | Blattulidae* | - - - - - - - - - - - - - - - - - - - - - - - - - |  |  |
|  | Caloblattinidae | - - - - - - - - - - - - - - - - - - - - - - - - - |  |  |

| Order | Family | Lebanese | Burmese | Canadian |
|-------|--------|----------|---------|----------|
| | | 145.5 mya | | 65.5 mya |
| | Mesoblattinidae* | - - - - - - - - - - - - - - - - - - - - - - - - - - - | | |
| | Polyphagidae | - - - - - - - - - - - B - - - - - - - - - - - - | | |
| | Raphidiomimidae | - - - - - - - - - - - B | | |
| | Umenocoleidae* | - - L - - - - - - - - - - - - - - - - - - - - | | |
| | undetermined | - - L - - - - - - - - B - - - - - - - - C - - - | | |
| **Dermaptera** | Labiduridae | - - L - - - - - - - - B - - - - - - - - - - - - | | |
| (earwigs) | Pygidicranidae | - - - - - - - - - - - B - - - - - - - - - - - - | | |
| **Zoraptera** | Zorotypidae | - - L - - - - - - - - B - - - - - - - - - - - - | | |
| (zorapterans) | | | | |
| **Paraplecoptera*** | Chresmodidae* | - - L - - - - - - - - - - - - - - - - - - - - | | |
| **Plecoptera** | Baleyopterygidae* | - - - - - - - - - - - - - - - - - - - - - - - - - - - - | | |
| (stoneflies) | Chloroperlidae | - - - - - - - - - - - - - - - - - - - - - - - - - - - - | | |
| | Gripopterygidae | - - - - - - - - - - - - - - - - - - - - - - - - - - - - | | |
| | Leuctridae | - - - - - - - - - - - - - - - - - - - - - - - - - - - - | | |
| | Nemouridae | - - - - - - - - - - - - - - - - - - - - - - - - - - - - | | |
| | Perlariopseidae* | - - - - - - - - - - - - | | |
| | Perlidae | | - - - - - - - - - - - - - - | |
| | Siberioperlidae* | - - - - - - - - - - - - | | |
| | Taeniopterygidae | - - - - - - - - - - - - - - - - - - - - - - - - - - - - | | |
| | undetermined | - - - - - - - - - - - B - - - - - - - - - - - - | | |
| **Embioptera** | Burmitembiidae* | - - - - - - - - - - - B | | |
| (webspinners) | | | | |
| **Isoptera** | Hodotermitidae | - - - - - - - - - - - B - - - - - - - - - - - - | | |
| (termites) | Kalotermitidae | - - L - - - - - - - - B - - - - - - - - - - - - | | |
| | Mastotermitidae | - - - - - - - - - - - - - - - - - - - - - - - - - - - - | | |
| | Rhinotermitidae | | B - - - - - - - - - - - - | |
| | Termitidae | - - - - - - - - - - - B - - - - - - - - - - - - | | |
| | Termopsidae | - - - - - - - - - - - - - - - - - - - - - - - - - - - - | | |

| Order | Family | Lebanese | Burmese | Canadian |
|-------|--------|----------|---------|----------|
| | | 145.5 mya | | 65.5 mya |
| **Mantida** | Amorphoscelidae | - - - - - - - - - - - - - - - - - - - - - - - - - - | | |
| (mantids) | Baissomantidae* | - - - - - - - | | |
| | Chaeteessidae | - - - - - - - - - - - - - - - - - - - - - - - - | | |
| | Cretomantidae* | - - - - - - - - - - - - - - - - - - - - - - - - | | |
| | undetermined | - - L - - - - - - - - B - - - - - - - - C - - - | | |
| | | | | |
| **Grylloblattida** | Blattogryllidae* | - - - - - - - - - - - - | | |
| (rock crawlers) | Geinitziidae* | - - - - - - - - - - - - | | |
| | Mesorthopteridae* | - - - - - - - - - - - - | | |
| | Oecanthoperlidae* | - - - - - - - - - - - - | | |
| | | | | |
| **Orthoptera** | Baissogryllidae* | - - - - - - - - - - - | | |
| (crickets, | Gryllidae | - - - - - - - - - - - B - - - - - - - - C - - - | | |
| grasshoppers) | Gryllotalpidae | - - - - - - - - - - - - - - - - - - - - - - - - - | | |
| | Elcanidae* | - - - - - - - - - - - B | | |
| | Eumastacidae | - - - - - - - - - - - - - - - - - - - - - - - - - - | | |
| | Haglidae | - - - - - - - - - - - - - - - - - - - - - - - - - | | |
| | Haglotettigoniidae* | - - - - - - | | |
| | Locustopsidae | - - - - - - - - - - - - - - - - - - - - - - - - - | | |
| | Phasmomimidae* | - - - - - - - - - - - - - - - - - - - - - - - - - | | |
| | Prophalangopsidae | - - - - - - - - - - - - - - - - - - - - - - - - - | | |
| | Tetrigidae | - - - - - - - - - - - B - - - - - - - - - - - - | | |
| | Tettigonidae | - - - - - - - - - - - B - - - - - - - - - - - - | | |
| | Tridactylidae | - - - - - - - - - - - B - - - - - - - - - - - - | | |
| | Stenopelmatidae | - - - - - - - - - - - - - - - - - - - - - - - - - | | |
| | Vitimiidae* | - - - - - - | | |
| | | | | |
| **Phasmatodea** | | | | |
| (walking sticks) | Aerophasmatidae* | - - - - - - - - - - - - - - - - - - - - - - - - - | | |
| | Cretophasmatidae* | - - - - - - - - - - - - - - - - - - - - - - - - - | | |
| | Susumaniidae* | - - - - - - - - - - - - - - - - - - - - - - - - - | | |
| | undetermined | - - - - - - - - - - - B - - - - - - - C - - - | | |
| | | | | |
| **Phthiriaptera** | Saurodectidae* | - - - - - - | | |

| Order | Family | Lebanese | Burmese | Canadian |
|-------|--------|----------|---------|----------|
| | | 145.5 mya | | 65.5 mya |
| **Hemiptera** | Achilidae | - - - - - - - - - - - - B - - - - - - - - - - - - - | | |
| (aphids, scales, | Albicoccidae* | B - - - - - - - - - - - - | | |
| bugs) | Aleyrodidae | - - L - - - - - - - - B - - - - - - - - - - - - | | |
| | Alydidae | - - - - - - - - - - - - - - - - - - - - - - - - - | | |
| | Anthocoridae | - - L - - - - - - - - - - - - - - - C - - - | | |
| | Aphididae | - - - - - - - - - - - - - - - - - - - - - C - - - | | |
| | Aphrophoridae | - - B - - - - - - - - - - - - | | |
| | Aradidae | - - - - - - - - - - - B - - - - - - - - - - - - | | |
| | Belostomatidae | - - - - - - - - - - - - - - - - - - - - - - - - - | | |
| | Biturritiidae | - - - - - - - - - - - - - - - - - - - - - - - - - | | |
| | Burmacoccidae* | B - - - - - - - - - - - - | | |
| | Burmitaphidae* | - - - - - - - - B | | |
| | Canadaphididae* | - - - - - - - - C - - - | | |
| | Cercopidae | - - - - - - - - - - - - C - - - | | |
| | Cercopionidae* | - - - - - | | |
| | Cicadellidae | - - - - - - - - - - - B - - - - - - - - - - - - | | |
| | Cicadidae | - - - - - - - - - - - - - - | | |
| | Cixiidae | - - L - - - - - - - - B - - - - - - - - C - - - | | |
| | Coreidae | - - - - - - - - - - - B - - - - - - - - - - - - | | |
| | Corixidae | - - - - - - - - - - - - - - - - - - - - - - - - - | | |
| | Cretamyzidae* | - - - - - - - - - - - C - - - | | |
| | Cydnidae | - - - - - - - - - - - B - - - - - - - - - - - - | | |
| | Dictyopharidae | - - - - - - - - - - - - - - - - - - - - - - - - - | | |
| | Dipsocoridae | - - L - - - - - - - - B - - - - - - - - - - - - | | |
| | Drepanosiphidae | - - - - - - - - - - - - - - - - - - - - - C - - - | | |
| | Electrococcidae* | - - L - - - - - - - - - - - - - - - - - - - - | | |
| | Elektraphididae* | - - - - - - - - - - - - - - - - - - - - - - - - - | | |
| | Enicocephalidae | - - L - - - - - - - - B - - - - - - - - - - - - | | |
| | Enicocoridae* | - - - - - - | | |
| | Eriococcidae | - - - - - - - - - - - - - - - - - - - - - - - - - | | |
| | Eurymelidae* | - - - - - - - - - - - - - - - - - - - - - - - - - | | |
| | Fulgoridae | - - L - - - - - - - - B - - - - - - - - - - - - | | |
| | Gerridae | - - - - - - - - | | |
| | Grimaldiellidae* | - - - - - - - | | |

| Order | Family | Lebanese | Burmese | Canadian |
|---|---|---|---|---|
| | | 145.5 mya | | 65.5 mya |
| | Hyarometridae* | - - - - - - - - - - - - - - - - - - - - - - - - - - |
| | Hydrometridae | | B - - - - - - - - - - - - |
| | Hylicellidae* | - - - - - - |
| | Inkaidae* | | - - - - - - - - - - - - - |
| | Jersicoccidae* | | - - - - - - - - - - - - - - - - - - |
| | Karabasiidae* | - - - - - - |
| | Karajassidae* | - - - - - - |
| | Kukaspididae* | - - - - - - |
| | Labiococcidae* | | | - - - - - - - |
| | Lalacidae* | - - - - - - - - - - - - - - - - - - - - - - - - - |
| | Liadopsyllidae* | - - - - - - |
| | Lophopidae | - - - - - - - - - - - - - - - - - - - - - - - - - |
| | Lygaeidae | | - - - - - - - - - - - - - - - - - - |
| | Margarodidae | - - - - - - - - - - - - - - - - - - - - C - - - |
| | Matsucoccidae | - - - - - - - - - - - - - - - - - - - - - - - - - |
| | Membracidae | - - - - - - - - - - - - - - - - - - - - - - - - - |
| | Mesopentacoridae* | - - - - - - |
| | Mesotrephidae* | | | - - - - - - - |
| | Mesoveliidae | - - - - - - - - - - - - - - - - - - - - - - - - - |
| | Mesozoicaphididae* | - - - - - - - - - - - - - - - - - - - - C - - - |
| | Mindaridae | | | - - C - - - |
| | Miridae | - - L - - - - - - - - - - - - - - - - - - - - |
| | Monophlebiidae | - - L - - - - - - - - - - - - - - - - - - - - |
| | Nabidae | - - - - - - - - - - - - - - - - - - - - - - - - - |
| | Naucoridae | - - - - - - - - - - - - - - - - - - - - - - - - - |
| | Notonectidae | - - - - - - - - - - - - - - - - - - - - - - - - - |
| | Ochteridae | - - - - - - - - - - - B - - - - - - - - - - - - |
| | Ortheziidae | - - L - - - - - - - - B - - - - - - - - - - - - |
| | Oviparosiphidae* | - - - - - - |
| | Pachymeridiidae* | - - - - - - |
| | Palaeoaphididae* | - - - - - - - - - - - - - - - - - - - - C - - - |
| | Palaeontinidae* | - - - - - - |
| | Parvaverrucosidae* | | - - B - - |
| | Pemphigidae | - - - - - - - - - - - - - - - - - - - - - - - - - |

| Order | Family | Lebanese | Burmese | Canadian |
|---|---|---|---|---|
| | | 145.5 mya | | 65.5 mya |
| | Pentatomidae | ------------------------- | | |
| | Phylloxeridae | --------------------- C --- | | |
| | Piesmatidae | | B ------------- | |
| | Procercopidae* | ------ | | |
| | Progonocimicidae* | ------ | | |
| | Protopsyllidiidae* | ------------ | | |
| | Pseudococcidae | | | -------- |
| | Psyllidae | ------------------------- | | |
| | Putoidae | - - L --------------------- | | |
| | Reduvidae | ------------ B -------- C --- | | |
| | Saldidae | ------------------------- | | |
| | Schizopteridae | ------------ B ------------- | | |
| | Shaposhnikoviidae* | ------ | | |
| | Steingeliidae | - - L --------------------- | | |
| | Tajmyraphididae* | - - L ------------------ C --- | | |
| | Thaumastellidae | - - L --------------------- | | |
| | Thaumastocoridae | ------------------------- | | |
| | Tingidae | ------------------------- | | |
| | Vianaididae | ------------------------- | | |
| | Xylococcidae | --------------------- C --- | | |
| **Psocoptera** | Amphientomidae | | ------------ | |
| (bark lice) | Archipsyllidae | ------------------------- | | |
| | Compsocidae | ------------ B ------------- | | |
| | Ellipsocidae | | ------------- | |
| | Lachesillidae | | ------------- | |
| | Liposcelidae | | B ------------- | |
| | Mesopsocidae | - -L --------------------- | | |
| | Pachytroctidae | ------------ B ------------- | | |
| | Prionoglariidae | - - L --------------------- | | |
| | Psocidae | | ------------ | |
| | Psyllipsocidae | ------------ B ------------- | | |
| | Sphaeropsocidae | - - L ------------------ C --- | | |
| | Trogiidae | | B ------------- | |
| | undetermined | - - L -------- B -------- C --- | | |

| Order | Family | Lebanese | Burmese | Canadian |
|---|---|:---:|:---:|:---:|
| | | 145.5 mya | | 65.5 mya |
| **Thysanoptera** (thrips) | Aeolothripidae | | B | |
| | Heterothripidae | L | | |
| | Jezzinothripidae* | L | | |
| | Lophioneuridae* | L | B | |
| | Neocomothripidae* | L | | |
| | Rhetinothripidae* | L | | |
| | Scapthothripidae* | L | | |
| | Scudderothripidae* | L | | |
| | Thripidae | | B | |
| | undetermined | L | B | C |
| **Neuroptera** (lacewings, ant lions, mantid flies) | Babinskaiidae* | | | |
| | Berothidae | L | B | C |
| | Chrysopidae | | | |
| | Coniopterygidae | L | B | |
| | Hemerobiidae | | | |
| | Kalligrammatidae* | | | |
| | Mantispidae | L | | |
| | Myrmeleontidae | L | | |
| | Nemopteridae | | | |
| | Nymphidae | | | |
| | Osmylidae | | B | |
| | Psychopsidae | | | |
| | Sisyridae | | | C |
| **Rhaphidioptera** (snake flies) | Alloraphidiidae* | | | |
| | Baissopteridae* | | | |
| | Mesoraphidiidae* | | B | |
| | Rhachiberothidae | L | | |
| | undetermined | | B | C |
| **Megaloptera** (alderflies, dobsonflies) | Corydalidae | | | |
| | Sialidae | | | |
| **Mecoptera** (scorpion flies) | Aneuretopsychidae* | | | |
| | Bittacidae | | | |
| | Boreidae | | | |

| Order | Family | Lebanese | Burmese | Canadian |
|-------|--------|----------|---------|----------|

145.5 mya               65.5 mya

| Order | Family | Lebanese | Burmese | Canadian |
|-------|--------|:--------:|:-------:|:--------:|
| | Englathaumatidae* | - - - - - - - - - - - - - - - - - - - - - - - - - | | |
| | Mesopsychidae* | - - - - - - - - - - - - - - - - - - - - - - - - - | | |
| | Nannochoristidae | | - - - - - - - - - - - - - | |
| | Orthophlebiidae | - - - - - - - - - - - - - - - - - - - - - - - - - | | |
| | Pseudopolycentro-<br>podidae* | - - - - - - - - - - - B | - - - - - - - - - - - - | |
| **Siphonaptera**<br>(fleas) | Saurophthiridae* | - - - - - - | | |
| **Coleoptera**<br>(beetles) | Acanthocnemidae | - - - - - - - - - - - - - - - - - - - - - - - - - - | | |
| | Ademosynidae* | - - - - - - - - - - - - - - - - - - - - - - - - - | | |
| | Aderidae | - - L - - - - - - - - B | - - - - - - - - - - - | |
| | Alleculidae | - - - - - - - - - - - - - - - - - - - - - - - - - | | |
| | Anthicidae | - - L - - - - - - - - B | - - - - - - - - - - - | |
| | Attelabidae | - - - - - - - - - - - - - - - - - - - - - - - - - | | |
| | Boganiidae | - - L - - - - - - - - - - - - - - - - - - | - - - - | |
| | Brentidae | | | - - - - - - - |
| | Bruchidae | | | - - - C - - - |
| | Buprestidae | | B - - - - - - - - - - - | - - |
| | Callirhypidae* | | | - - - - - - - |
| | Cantharidae | | B - - - - - - - - - - - | - - |
| | Carabidae | - - L - - - - - - - - B | - - - - - - - - - - - | |
| | Caridae | - - - - - - - - - - - - - - - - - - - - - - - - - | | |
| | Cerambycidae | | B - - - - - - - - - - - | - - |
| | Cerophytidae | - - - - - - - - - - - - - - - - - - - - - - - - - | | |
| | Chrysomelidae | - - - - - - - - - - - B | - - - - - - - - - - | |
| | Cisidae | - - - - - - - - - - - B | - - - - - - - - - - | |
| | Cleridae | - - - - - - - - - - - B | - - - - - - - - - - | |
| | Coccinellidae | | | - - - - - - - |
| | Colydiidae | - - L - - - - - - - - B | - - - - - - - - - - - | |
| | Coptoclavidae* | - - - - - - - - - - - | | |
| | Corylophidae | | | - - - - - - - |
| | Cryptophagidae | - - - - - - - - - - - - - - - - - - - - - - - - - | | |
| | Cucujidae | - - L - - - - - - - - B | - - - - - - - - - - - | |

| Order | Family | Lebanese | Burmese | Canadian |
|---|---|---|---|---|
| | | 145.5 mya | | 65.5 mya |
| | Cupedidae | -------------------------- | | |
| | Curculionidae | -------------------------- | | |
| | Dascillidae | -------------------------- | | |
| | Dermestidae | - - L - - - - - - - - B - - - - - - - - - - - - | | |
| | Dytiscidae | -------------------------- | | |
| | Eccoptarthridae* | | B - - - - - - - - - - - - - | |
| | Elateridae | - - L - - - - - - - - B - - - - - - - - - - - - | | |
| | Endomychidae | - - L - - - - - - - - B - - - - - - - - - - - - | | |
| | Eucinetidae | - - - - - - - - - - - - B - - - - - - - - - - - - | | |
| | Eucmetidae | - - - - - - - - - - - - - - - - - - - - - C - - - | | |
| | Eucnemidae | - - - - - - - - - - - - B - - - - - - - - - - - - | | |
| | Gyrinidae | -------------------------- | | |
| | Haplochelidae* | | - - B - - | |
| | Helodidae | - - - - - - - - - - - - B - - - - - - - - C - - - | | |
| | Histeridae | -------------------------- | | |
| | Hydrophilidae | -------------------------- | | |
| | Labradorocoleidae* | -------------------------- | | |
| | Lathridiidae | - - L - - - - - - - - B - - - - - - - - - - - - | | |
| | Leiodidae | -------------------------- | | |
| | Lymexylidae | - - - - - - - - - - - - B - - - - - - - - - - - - | | |
| | Melandriidae | | | - - - - - - - - - - |
| | Melyridae | -------------------------- | | |
| | Micromalthidae | - - L - - - - - - - - - - - - - - - - - - - - - | | |
| | Microsporidae | - - - - - - - - - - - - B - - - - - - - - - - - - | | |
| | Mordellidae | - - L - - - - - - - - B - - - - - - - - C - - - | | |
| | Mycetophagidae | - - L - - - - - - - - - - - - - - - - - - C - - - | | |
| | Nemonychidae | - - L - - - - - - - - - - - - - - - - - - - - - | | |
| | Nitidulidae | - - - - - - - - - - - - B - - - - - - - - - - - - | | |
| | Oedemeridae | - - - - - - - - - - - - B - - - - - - - - - - - - | | |
| | Parandrexidae* | -------------------------- | | |
| | Platypodidae | - - L - - - - - - - - - - - - - - - - - - - - - | | |
| | Pselaphidae | - - L - - - - - - - - B - - - - - - - - - - - - | | |
| | Ptiliidae | - - L - - - - - - - - B - - - - - - - - - - - - | | |
| | Ptilodactylidae | - - - - - - - - - - - - B - - - - - - - - - - - - | | |

| Order | Family | Lebanese | Burmese | Canadian |
|-------|--------|----------|---------|----------|
| | | 145.5 mya | | 65.5 mya |
| | Prostomidae | | - - - - - - - - - - - B - - - - - - - - - - - - - | |
| | Rhipiphoridae | | - - - - - - - - - - - B - - - - - - - - - - - - - | |
| | Salpingidae | - - L - - - - - - - B - - - - - - - - - - - - - | | |
| | Scarabaeidae | - - L - - - - - - - B - - - - - - - - - - - - - | | |
| | Scirtidae | | | - - - - - - - - - - - - - |
| | Scolytidae | - - L - - - - - - - B - - - - - - - - - - - - - | | |
| | Scraptiidae | | - - - - - - - - - - - B - - - - - - - - - - - - - | |
| | Scydmaenidae | - - L - - - - - - - B - - - - - - - - C - - - | | |
| | Silvanidae | | - - - - - - - - - - - B - - - - - - - - - - - - - | |
| | Staphylinidae | - - L - - - - - - - B - - - - - - - - C - - - | | |
| | Tenebrionidae | - - - - - - - - - - - - - - - - - - - - - - - - - | | |
| | Throscidae | | - - - - - - - - - - - B - - - - - - - - - - - - - | |
| | Trachypachidae | - - - - - - - - - - - - - - - - - - - - - - - - - | | |
| | Trogossitidae | - - L - - - - - - - B - - - - - - - - - - - - - | | |
| | Ulyanidae* | - - - - - - - - - - | | |
| **Strepsiptera** | Mengeidae? | | B - - - - - - - - - - - - - | |
| **Diptera** | Acroceridae | | - - - - - - - - - - - B - - - - - - - - - - - - - | |
| (flies) | Agromyzidae | | | - - - - - - - - - - - - - |
| | Anisopodidae | - - L - - - - - - - B - - - - - - - - C - - - | | |
| | Antefungivoridae* | - - - - - - - - - - - - - - - - - - - - - - - - - | | |
| | Apioceridae | - - - - - - - - - - - - - - - - - - - - - - - - - | | |
| | Apsilocephalidae | | - - - - - - - - - - - B - - - - - - - - - - - - - | |
| | Archisargidae* | - - - - - - | | |
| | Archizelmiridae* | | - - - - - - - - - - - B - - - - - - - - - - - - - | |
| | Asilidae | - - L - - - - - - - - - - - - - - - - - - - - | | |
| | Atelestidae | - - L - - - - - - - - - - - - - - - - - - - - | | |
| | Athericidae | - - - - - - - - - - - - - - - - - - - - - - - - - | | |
| | Axymyiidae | - - - - - - - - - - - - - - - - - - - - - - - - - | | |
| | Bibionidae | - - L - - - - - - - - - - - - - - - - - C - - - | | |
| | Blephariceridae | | - - - - - - - - - - - B - - - - - - - - - - - - - | |
| | Boholdoyidae* | - - - - - - | | |
| | Bolitophilidae | - - - - - - - - - - - - - | | |

| Order | Family | Lebanese | Burmese | Canadian |
|-------|--------|----------|---------|----------|

145.5 mya                                          65.5 mya

| Family | | |
|--------|---|---|
| Bombyliidae | - - - - - - - - - - - - - - - - - - - - - - - - - |
| Calliphoridae | - - - - - - - |
| Cecidomyiidae | - - L - - - - - - - - B - - - - - - - - C - - - |
| Ceratopogonidae | - - L - - - - - - - - B - - - - - - - - C - - - |
| Chaoboridae | - - L - - - - - - - - B - - - - - - - - - - - - - |
| Chironomidae | - - L - - - - - - - - B - - - - - - - C - - - |
| Chloropidae | - - L - - - - - - - - - - - - - - - - - - C - - - |
| Corethrellidae | - - L - - - - - - - - B - - - - - - - - - - - - - |
| Cramptonomyiidae | - - - - - - - - - - - - - - |
| Culicidae | - - - - - - - - - - - - - - - - - - - - C - - - |
| Diadocidiidae | - - - - - - - - - - - - B - - - - - - - - - - - - |
| Dolichopodidae | - - L - - - - - - - - B - - - - - - - - C - - - |
| Elliidae* | - - - - - - |
| Empididae | - - L - - - - - - - - B - - - - - - - - C - - - |
| Eoditomyiidae* | - - - - - - |
| Eoptychopteridae* | - - L - - - - - - - - - - - - - - - - - - - - |
| Eremochaetidae* | - - - - - - - - - - - - - - - - - - - - - - - - |
| Hilarimorphidae | - - - - - - - - - - - - B - - - - - - - - - - - - |
| Hybotidae | - - - - - - - - - - - - - - - - - - - - - - - - |
| Ironomyiidae | - - L - - - - - - - - - - - - - - - - - - C - - - |
| Keroplatidae | - - L - - - - - - - - B - - - - - - - - - - - - - |
| Limoniidae | - - L - - - - - - - - B - - - - - - - - - - - - - |
| Lonchopteridae | - - L - - - - - - - - - - - - - - - - - - - - - |
| Mesosciophilidae* | - - - - - - - - - - - - - - - - - - - - - - - - |
| Milichiidae | - - - - - - - - - - - - - - - - - - - - - - - - |
| Mycetophilidae | - - L - - - - - - - - B - - - - - - - - C - - - |
| Nemestrinidae | - - - - - - - - - - - - - - - - - - - - - - - - |
| Opetiidae | - - - - - - - - - - - - - - - - - - - - - - - - |
| Pachyneuridae | - - - - - - - - - - - - - - - - - - - - - - - - |
| Paraxymyiidae* | - - - - - - - - - - - - - - - - - - - - - - - - |
| Perissommatidae | - - - - - - - - - - - - - - - - - - - - - - - - |
| Phlebotomidae | - - L - - - - - - - - B - - - - - - - - - - - - - |
| Phoridae | - - L - - - - - - - - B - - - - - - - - C - - - |
| Pipunculidae | - - - - - - - - - - - - - - - - - - - - C - - - |

| Order | Family | Lebanese | Burmese | Canadian |
|-------|--------|----------|---------|----------|
| | | 145.5 mya | | 65.5 mya |
| | Platypezidae | ----------------------C --- | | |
| | Pleciofungivoridae* | ------------------------- | | |
| | Procramptonomyiidae* | ------------------------- | | |
| | Protorhyphidae* | ----------- | | |
| | Psychodidae | - - L - - - - - - - - B - - - - - - - - - - - - - | | |
| | Ptychopteridae | ------------------------- | | |
| | Rhagionempididae* | ------------------------- | | |
| | Rhagionidae | - - L - - - - - - - - B - - - - - - - - - - - - - | | |
| | Scatopsidae | ----------- B --------C --- | | |
| | Sciadoceridae | - - L - - - - - - - - - - - - - - - -C --- | | |
| | Sciaridae | - - L - - - - - - - - B - - - - - - - -C --- | | |
| | Simuliidae | ------------------------- | | |
| | Stratiomyidae | ----------------------C --- | | |
| | Synneuridae | ----------------------C --- | | |
| | Tabanidae | ------------------------- | | |
| | Tanyderidae | - - L - - - - - - - - B - - - - - - - - - - - - - | | |
| | Thaumaleidae | ------------------------- | | |
| | Therevidae | ----------- B - - - - - - - - - - - - - | | |
| | Tipulidae | - - L - - - - - - - - B - - - - - - - -C --- | | |
| | Trichoceridae | - - L - - - - - - - - - - - - - - - - - - - - | | |
| | Xylophagidae | ------------------------- | | |
| | Vermileonidae | ------------------------- | | |
| **Hymenoptera** (wasps, bees, ants) | Ampulicidae | - - L - - - - - - - - B - - - - - - - - - - - - - | | |
| | Anaxyelidae | ------------------------- | | |
| | Andreneliidae* | - - - - - - | | |
| | Angarosphecidae* | ------------------------- | | |
| | Archaeocynipidae* | - - - - - - - - | | |
| | Aulacidae | ------------------------- | | |
| | Austroniidae | ------------------------- | | |
| | Baissodidae* | ------------------------- | | |
| | Bethylidae | - - L - - - - - - - - B - - - - - - - - - - - - - | | |
| | Bethylonymidae* | - - - - - - - - | | |

| Order | Family | Lebanese | Burmese | Canadian |
|---|---|---|---|---|
| | | 145.5 mya | | 65.5 mya |
| | Braconidae | - - - - - - - - - - - - B | - - - - - - - - C - - - | |
| | Cephidae | - - - - - - - - - - - - - - - - - - - - - - - - - | | |
| | Ceraphronidae | - - L - - - - - - - - - - - - - - - - - | - C - - - | |
| | Chalcididae | - - L - - - - - - - B | - - - - - - - - C - - - | |
| | Chrysididae | - - L - - - - - - - - - - - - - - - - - - - | | |
| | Cretevaniidae* | - - - - - - - - - - - - - - - - - - - - - - - - - | | |
| | Cynipidae | | - - - - - - - - C - - - | |
| | Diapriidae | - - - - - - - - - - - - B - - - - - - - - - - - - | | |
| | Dryinidae | - - L - - - - - - - B | - - - - - - - C - - - | |
| | Embolemidae | - - - - - - - - - - - - B - - - - - - - - - - - | | |
| | Eoichneumonidae* | - - - - - - | | |
| | Ephialtitidae* | - - - - - - - - - - - - - - - - - - - - - - - - - | | |
| | Eulophidae | - - - - - - - - - - - - - - - - - - - - - C - - - | | |
| | Eupelmidae | - - - - - - - - - - - - - - - - - - - - - C - - - | | |
| | Evaniidae | - - L - - - - - - - B | - - - - - - - - - - - - | |
| | Falsiformicidae* | - - - - - - - - - - - - B | | |
| | Figitidae | | - - - - - - - - - - - - - - | |
| | Formicidae | B | - - - - - - - - C - - - | |
| | Gasteruptiidae | - - L - - - - - - - B | - - - - - - - - - - - - | |
| | Heloridae | - - - - - - - - - - - - - - - - - - - - - C - - - | | |
| | Ichneumonidae | - - - - - - - - - - - - - - - - - - - - - C - - - | | |
| | Ichneumonomimidae* | - - - - - - | | |
| | Maimetshidae* | - - - - - - - - - - - - - - - - - - - - - - - - - | | |
| | Megalodontidae | - - - - - - - - - - - - - - - - - - - - - - - - - | | |
| | Megalyridae | - - L - - - - - - - B | - - - - - - - - - - - - | |
| | Megaspilidae | - - L - - - - - - - B | - - - - - - - - - - - - | |
| | Melittosphecidae* | | - - B - - | |
| | Mesoserphidae* | - - - - - - | | |
| | Monomachidae | - - - - - - - - - - - - - - - - - - - - - - - - - | | |
| | Mutillidae | - - - - - - - - - - - - - - - - - - - - - - - - - | | |
| | Mymaridae | - - - - - - - - - - - - B | - - - - - - - - C - - - | |
| | Mymarommatidae | - - L - - - - - - - B | - - - - - - - - - - - - | |
| | Orussidae | - - - - - - - - - - - - - - - - - - - - - - - - - | | |
| | Pamphiliidae | - - - - - - - - - - - - - - - - - - - - - - - - - | | |
| | Pelecinidae | - - - - - - - - - - - - B - - - - - - - - - - - | | |

| Order | Family | Lebanese | Burmese | Canadian |
|-------|--------|----------|---------|----------|

145.5 mya               65.5 mya

| Order | Family | Lebanese | Burmese | Canadian |
|-------|--------|----------|---------|----------|
| | Platygastridae | | - - - - - - - - - - - - - - - - - | |
| | Plumariidae | - - - - - - - - - - - - - - - - - - - - - - - - | | |
| | Pompilidae | - - - - - - - - - - - B - - - - - - - - - - - - | | |
| | Praeaulacidae* | - - - - - - | | |
| | Praeichneumonidae* | - - - - - - | | |
| | Praesiricidae* | - - - - - - - - - - - - - - - - - - - - - - - - | | |
| | Proctotrupidae | - - L - - - - - - - - - - - - - - - - - C - - - | | |
| | Pseudosiricidae* | - - - - - - - - - - - - - - - - - - - - - - - - | | |
| | Rasnicynipidae* | - - - - - - | | |
| | Rhopalosomatidae | - - - - - - - - - - - - - - - - - - - - - - - - | | |
| | Roproniidae | - - - - - - - - - - - - - - - - - - - - - - - - | | |
| | Scelionidae | - - L - - - - - - - B - - - - - - - - C - - - | | |
| | Scoliidae | - - - - - - - - - - - - - - - - - - - - - - - - | | |
| | Scolobythidae | - - L - - - - - - - - - - - - - - - - - - - | | |
| | Sepulcidae* | - - - - - - | | |
| | Serphitidae* | - - L - - - - - - - B - - - - - - - - C - - - | | |
| | Sierolomorphidae | - - - - - - - - - - - B - - - - - - - - - - - - | | |
| | Siricidae | - - - - - - - - - - - - - - - - - - - - - - - - | | |
| | Sphecidae | - - L - - - - - - - B - - - - - - - C - - - | | |
| | Stigmaphronidae | - - L - - - - - - - B - - - - - - - C - - - | | |
| | Tenthredinidae | - - - - - - - - - - - - - - - - - - - - - - - - | | |
| | Tetracampidae | - - - - - - - - - - - - - - - - - - - - - C - - - | | |
| | Tiphiidae | - - - - - - - - - - - B - - - - - - - - - - - - | | |
| | Torymidae | - - - - - - - - - - - - - - - - - - - - - C - - - | | |
| | Trichogrammatidae | - - - - - - - - - - - - - - - - - - - - - - - - | | |
| | Trigonalidae | - - - - - - - - - - - - - - - - - - - - - - - - | | |
| | Vanhorniidae | - - - - - - - - - - - - - - - - - - - - - - - - | | |
| | Vespidae | - - - - - - - - - - - - - - - - - - - - - - - - | | |
| | Xyelotomidae* | - - - - - - - - | | |
| | Xyelydidae* | - - - - - - | | |
| **Polyneoptera** | Chresmodidae* | - - - - - - | | |
| **Trichoptera** (caddis flies) | Baissoferidae* | - - - - - - - - - - - - | | |
| | Brachycentridae | - - - - - - - - - - - - - - - - - - - - - - - - - | | |
| | Calamoceratidae* | | | - - - - - - - |

| Order | Family | Lebanese | Burmese | Canadian |
|-------|--------|----------|---------|----------|
| | | 145.5 mya | | 65.5 mya |
| | Dysoneuridae* | - - - - - - - - - - | | |
| | Electralbertidae | - - - - - - - - - - - - - - - - - - - - - C - - - | | |
| | Helicophidae | - - - - - - - - - - - - - - - - - - - - - - - - | | |
| | Hydrobiosidae | - - - - - - - - - - - - - - - - - - - - - - - - | | |
| | Hydroptilidae | - - - - - - - - - - - B - - - - - - - - - - - - - | | |
| | Lepidostomatidae | - - - - - - - - - - - - - - - - - - - - - - - - - | | |
| | Leptoceridae | - - - - - - - - - - - - - - - - - - - - - - - - - | | |
| | Necrotauliidae* | - - - - - - - - - - | | |
| | Odontoceridae | | | - - - - - - - |
| | Philopotamidae | - - - - - - - - - - - B - - - - - - - - - - - - - | | |
| | Phryganeidae | - - - - - - - - - - - - - - - - - - - - - - - - - | | |
| | Plectrotarsidae | - - - - - - - - - - - - - - - - - - - - - - - - - | | |
| | Polycentropodidae | - - - - - - - - - - - B - - - - - - - - - - - - - | | |
| | Rhyacophilidae | - - - - - - - - - - - - - - - - - - - - - - - - - | | |
| | Taimyrelectronidae* | - - - - - - - | | |
| | Vitimotauliidae* | - - - - - - - - - - - - - - - - - - - - | | |
| **Lepidoptera** | Bucculatricidae* | | | - - - - - - - |
| (moths) | Eolepidopterigidae* | - - L - - - - - | | |
| | Gelechiidae | | B - - - - - - - - - - - - | |
| | Gracillariidae | | B - - - - - - - - - - - - | |
| | Hepialidae | | - - - - - - - - - - - - - - - - - - - - | |
| | Incurvariidae | - - L - - - - - - - - - - - - - - - - - - - - - - | | |
| | Lophocoronidae* | | | - - - - - - - |
| | Micropterigidae | - - L - - - - - - - - B - - - - - - - - - - - - - | | |
| | Necrotauliidae* | - - - - - - - - - - - - - - - - - - - - - - - - - | | |
| | Nepticulidae | | | - - - - - - - - - - - - - |
| | Noctuidae | | | - - - - - - - |
| | Tineidae | - - - - - - - - - - - - - - - - - - - - - - - - - | | |
| | Undopterigidae* | - - L - - - - - - - - - - - - - - - - - - - - - - | | |

# APPENDIX B

## Key Factors Contributing to the Survival of Terrestrial Animals

CHARACTERISTICS that apply to most Hexapoda (six-legged arthropods) are in italics. Characteristics that may apply to most dinosaurs are in bold. Individual groups would have to be classified independently (exceptions exist). The more positive characteristics a group possesses, the less likely it will become extinct. $r$ = $r$-strategists; $K$ = $K$-strategists.

| Factors | Positive characteristics | Negative characteristics |
|---|---|---|
| Lineage | *Polyphyletic* | Monophyletic |
| Reproductive adaptations | | |
|   Reproductive rate | *High (r)* | Low *(K)* |
|   Onset of reproduction | *Early (r)* | **Late** *(K)* |
|   Generation time | *Short* | **Long** |
|   Broods/year | *Multiple (multivoltine)* | **Single (univoltine)** |
|   Reproductive cycles/lifetime | *Multiple* | Single |
|   Number of progeny/lifetime | *Many (r)* | Few *(K)* |
|   Newborn size:adult size ratio | *high* | **low** |
|   Growth rate | *Rapid (r)* | **Slow** *(K)* |
|   Progeny survival | High | Low |
|   Production of young | Synchronous | Asynchronous |
|   Neonatal care | *None (r)* | Some *(K)* |
|   Mate selection | Polygamy | Monogamy |
|   Breeding type | Outbreeding (exogamy) | Inbreeding (endogamy) |
|   Breeding sites | Multiple | Fixed, limited, or specialized |

| Factors | Positive characteristics | Negative characteristics |
|---|---|---|
| Adult characteristics | | |
| Body size | *Small (r)* | **Large** (K) |
| Population size | *Large* | Small |
| Biodiversity | *High* | **Low** |
| Distribution | *Global* | Endemic |
| Habitat type | *Varied (eurytopic)* | **Limited (stenotopic)** |
| Environment | *Unpredictable (r)* | **Stable** (K) |
| Adaptability | *Good (resilient) (r)* | **Poor** (K) |
| Dispersal | Extremely mobile (r) | Restricted (K) |
| Migratory patterns | *None or variable* | Fixed, inflexible |
| Social behavior | Solitary | Gregarious |
| Feeding type | Generalist | Specialist |
| Trophic level | Primary | High |
| Longevity | *Short* | **Long** |
| Mortality | *Density-independent (r)* | **Density-related** (K) |
| Environmental resources | *Rarely limited (r)* | Frequently limited (K) |
| Resistance to toxins | *High* | Low |
| Mutation rate | *High* | Low |
| Seasonal activity | *Resting phases* | Active year round |
| Temperature tolerance | *Broad* | Narrow |
| Predation risk | Low | High |
| Metabolic rate | *Low* | High |
| Body temperature | *Cold-blooded* | Warm-blooded |

---

## Problems with Evaluating the Fossil Record and Extinctions

**A.** Data base is too small.

**1.** Fossil record is incomplete. The vast majority of past organisms in the biosphere were rarely preserved.

**2.** Fossilization process is selective, favoring certain circumstances and habitats. Inclusions in amber are biased in that they selected small organisms that frequent habitats associated with resin-producing trees. Permineralized fossils occurred in areas where sediment could have inundated them. Large bones were more apt to be preserved than small ones. Organisms with hard structures such as shells and teeth are more frequently represented than soft-bodied creatures.

**3.** Fossil record is not continuous, leaving large gaps and often resulting in the sudden appearance of lineages previously thought to be extinct (Lazarus effect). Very few sites are currently known to span the K/T boundary, and even these have discontinuities.

**4.** At fossil sites, the absence of an organism does not mean that it was not present. Nor does the presence of a fossil tell you more than that a lineage was present at a particular time and place. A single fossil provides no information about the global distribution, how long the lineage survived, or when it became extinct. The apparent absence from one location does not preclude survival elsewhere. Accordingly, it is impossible to evaluate extinctions with negative data.

**5.** Cretaceous fossil sites are not distributed uniformly, so determining global patterns of various lineages is very difficult.

    **6.** Rare species may not be represented at all at some sites, but that doesn't mean they were not present.

**B.** Data may be flawed.

    **1.** Dating of fossils is often indirect, so their ages and the actual timing of some extinctions may be erroneous.

    **2.** Many fossils are redeposited from older strata into younger beds, thereby confusing the record (Zombie effect).

    **3.** Some uncertainties with species/genera identifications exist since most fossils are fragmentary and diagnostic characters are obscured or absent. Also, there are systematic inconsistencies among taxa of different organisms.

    **4.** As you approach extinction boundaries, the volume of rock available for sampling often decreases, thus lowering the chances of finding a particular fossil (Signor-Lipps effect).

**C.** Human factors.

    **1.** There appear to be strong biases operating in the identification of fossils and the evaluation of extinctions, and with such a small database, interpretations can be motivated by prejudice.

    **2.** Because there is no immediate urgency to resolve the issue of K/T extinctions, effort should be made to determine whether extinctions of various groups were abrupt or gradual, rather than supporting a cause.

**D.** Inherent scientific difficulties.

    **1.** Interpretation of fossils is based on inferences gathered from scientific knowledge representing only a few hundred years of human experience. Mistakes are probable because our current database may be flawed and is certainly incomplete.

    **2.** Predicting causes and effects of extinctions that occur today is often impossible. Extending predictions back 65 mya is even more problematic.

    **3.** A seemingly small ecological change, such as a slight global warming or cooling, could have far-reaching biological consequences both regionally and globally, while a dramatic event such as a meteor impact could be significantly less important because of the unpredictability of its consequences.

    **4.** Extinctions may be random, but random events can occur in clusters.

    **5.** Species have different degrees of importance in biological systems. The extinction of a keystone species may have a cascade effect and result in secondary extinctions that are a direct corollary but unrelated to the cause of the original extinction.

**6.** Species longevity differs. Species with short longevity spans that may have become extinct under normal circumstances cannot be distinguished from those that may have survived longer if an extinction event had not occurred. These short-lived species skew the interpretation of extinctions.

# References

1. Resh, V. H. & Cardé, R. T. 2003. Insecta. Overview, pp. 564–566 in Encyclopedia of Insects, Resh, V. H. & R. T. Cardé (eds.). Academic Press, Amsterdam.
2. Boucot, A. 1990. Evolutionary Paleobiology of Behavior and Co-evolution. Elsevier, Amsterdam, 725 pp.
3. Blaxter, M., Dorris, M. & De Ley, P. 2000. Patterns and processes in the evolution of animal parasitic nematodes. Nematology 2: 43–55.
4. Strong, D. R., Lawton, J. H. & Southwood, T. R. E. 1984. Insects on Plants: Community Patterns and Mechanisms. Harvard University Press, Cambridge, MA., 313 pp.
5. Labandeira, C. C. 2002. The history of associations between plants and animals. pp. 26–74 in Plant-Animal Interactions, Herrera, C. M. & Pellmyr, O. (eds.). Blackwell Science, Oxford.
6. Gradstein, F. M., Ogg, J. G. & Smith, A. G. (eds.). 2004. A Geologic Time Scale. Cambridge University Press, Cambridge, 589 pp.
7. Russell, D. A. 2000. The mass extinctions of the Late Mesozoic. pp. 370–380 in The Scientific American Book of Dinosaurs, Gregory, P. S., (ed.). St. Martin's Griffin, New York.
8. Webster, D. 1999. A dinosaur named Sue. National Geographic Magazine 195: 46–60.
9. Smith, A. G., Smith, D. G. & Funnel, B. M. 1994. Atlas of Mesozoic and Cenozoic Coastlines. Cambridge University Press, Cambridge, 99 pp.
10. Vakhrameev, V. A. 1988. Jurassic and Cretaceous Floras and Climates of the Earth. Cambridge University Press, Cambridge, 318 pp.
11. Azar, D. 2000. Les Ambres Mésozoïques du Liban. Ph.D. thesis, Univeristé de Paris, 164 pp.
12. Edwards, W. N. 1929. Lower Cretaceous plants from Syria and Transjordania. Annals and Magazine of Natural History 4: 394–405.

13. Poinar, Jr., G. O. & Milki, R. 2001. Lebanese Amber. Oregon State University Press, Corvallis, 96 pp.
14. Davies, E. H. 2001. Palynological analysis of two Burmese amber samples. Unpublished report by Branta Biostratigraphy Ltd. for Leeward Capital Corp., 6 pp.
15. Davies, E. H. 2001. Palynological analysis and age assignments of two Burmese amber sample sets. Unpublished report by Branta Biostratigraphy Ltd. for Leeward Capital Corp., 4 pp.
16. Grimaldi, D. A., Engel, M. S. & Nascimbene, P. C. 2002. Fossiliferous Cretaceous amber from Myanmar (Burma). American Museum Novitates 3361, 71 pp.
17. Poinar, Jr., G. O. 2004. *Programinis burmitis* gen. et sp. nov., and *P. laminatus* sp. nov., Early Cretaceous grass-like monocots in Burmese amber. Australian Systematic Botany 17: 497–504.
18. Poinar, Jr., G. O. & Chambers, K. 2005. *Palaeoanthella huangii* gen. and sp. nov., an Early Cretaceous flower (Angiospermae) in Burmese amber. Sida 21: 2087–2092.
19. Santiago-Blay, J. A., Anderson, S. R. & Buckley, R. T. 2005. Possible implications of two new angiosperm flowers from Burmese amber (Lower Cretaceous) for well-established and diversified insect-plant associations. Entomological News 116: 341–346.
20. Zherikhin, V. V. & Ross, A. J. 2000. A review of the history, geology and age of Burmese amber (Burmite). Bulletin of the Natural History Museum, London (Geology) 56: 3–10.
21. Braman, D. & Koppelhus, E. B. 2005. Campanian Palynomorphs. pp. 101–130 in Dinosaur Provincial Park, Currie, P. J. & Koppelhus, E. B. (eds.). Indiana University Press, Bloomington.
22. Jarzen, D. M. 1982. Palynology of Dinosaur Provincial Park (Campanion) Alberta. Syllogeus 38: 1–69.
23. Knowlton, F. H. 1905. Fossil plants of the Judith River beds. United States Geological Survey Bulletin 257:129–155.
24. Bell, W. A. 1965. Upper Cretaceous and Paleocene plants of Western Canada. Geological Survey of Canada Paper 65-35:1–46.
25. Colbert, E. H. 1983. Dinosaurs: An Illustrated History. Hammond Incorporated, Maplewood, NJ, 224 pp.
26. Poinar, Jr., G. O. 2005. A Cretaceous palm bruchid, *Mesopachymerus antiqua*, n. gen., n. sp. (Coleoptera: Bruchidae: Pachymerini) and biogeographical implications. Proceedings of the Entomological Society of Washington 107: 392–397.

27. Ostrom, J. H. 1964. A reconsideration of the paleoecology of hadrosaurian dinosaurs. American Journal of Science 262: 975–997.

28. Dodson, P. 1990. Counting dinosaurs: How many kinds were there? Proceedings of the National Academy of Sciences 87: 7608–7612.

29. Dodson, P. 1997. Distribution and diversity. pp. 186–188 in Encyclopedia of Dinosaurs, Currie, P. J. & Padian, K. (eds.). Academic Press, New York.

30. Fastovsky, D. E. & Weishampel, D. B. 2005. The Evolution and Extinction of the Dinosaurs, 2nd ed. Cambridge University Press, Cambridge, 485 pp.

31. Norman, D. 2000. The evolution of Mesozoic flora and fauna. pp. 204–230 in The Scientific American Book of Dinosaurs, Paul, G. S. (ed.). St. Martin's Griffin, New York.

32. Holtz, Jr., T. R., Chapman, R. E. & Lamanna, M. C. 2004. Mesozoic biogeography of Dinosauria. pp. 627–642 in The Dinosauria, 2nd ed., Weishampel, D. B., Dodson, P. & Osmólska, H. (eds.). University of California Press, Berkeley.

33. Daly, H. V., Doyen, J. T., & Purcell, A. H. 1998. Introduction to Insect Biology and Diversity, 2nd ed. Oxford University Press, Oxford, 680 pp.

34. Kuschel, G. & Poinar, Jr., G. O. 1993. *Libanorhinus succinus* gen. and sp. n. (Coleoptera: Nemonychidae) from Lebanese amber. Entomologica Scandanavica 24: 143–146.

35. Rasnitsyn, A. P. & Quicke, D. L. J. 2002. History of Insects. Kluver Academic Publishers, Dordrecht, 517 pp.

36. Whalley, P. 1978. New taxa of fossil and recent Micropterigidae with a discussion of their evolution and a comment on the evolution of Lepidoptera (Insecta). Annals of the Transvaal Museum 31: 65–81.

37. Zur Strassen, R. 1973. Fossile Fransenflugler aus mesozoischem Bernstein des Lebanon. Stuttgarter Beitrage für Naturkunde (Serie A) 267: 1–51.

38. Heie, O. & Azar, D. 2000. Two new species of aphids found in Lebanese amber and a revision of the family Tajmyaphididae Kononova, 1975 (Hemiptera: Sternorrhyncha). Annals of the Entomological Society of America 93: 1222–1225.

39. Poinar, Jr., G. O. & Brown, A. E. 2004. A new subfamily of Creta-

ceous antlike stone beetles (Coleoptera: Scydmaenidae: Hapsomelinae) with an extra leg segment. Proceedings of the Entomological Society of Washington 106: 789–796.

40. Kirejtshuk, A. G. & Poinar, Jr., G. O. 2006. Haplochelidae, a new family of Cretaceous beetles (Coleoptera, Myxophaga) from Burmese amber. Proceedings of the Entomological Society of Washington 108: 155–164.

41. Poinar, Jr., G. O. 2006. *Mesophyletis calhouni* (Mesophyletinae), a new genus, species and subfamily of Early Cretaceous weevils (Coleoptera: Curculionoidea: Eccoptarthridae) in Burmese amber. Proceedings of the Entomological Society of Washington 108: 878–884.

42. Prasad, V., Strömberg, C. A. E., Alimohammadian, H. & Sahni, A. 2005. Dinosaur coprolites and the early evolution of grasses and grazers. Science 310: 1177–1180.

43. Poinar, Jr., G. O. & Brown, A. 2005. New Aphidoidea (Hemiptera: Sternorrhyncha) in Burmese amber. Proceedings of the Entomological Society of Washington 107: 835–845.

44. Poinar, Jr., G. & Danforth, B. N. 2006. A fossil bee from Early Cretaceous Burmese amber. Science 314: 614.

45. Poinar, Jr., G. & Brown, A. E. 2002. A new genus of hard ticks in Cretaceous Burmese amber (Acari: Ixodida: Ixodidae). Systematic Parasitology 54: 199–205.

46. Poinar, Jr., G. O. & Brown, A. E. 2004. A new genus of primitive crane flies (Diptera: Tanyderidae) in Cretaceous Burmese amber, with a summary of fossil tanyderids. Proceedings of the Entomological Society of Washington 106: 339–345.

47. Poinar, Jr., G. O. & Brown, A. E. 2006. The enigmatic *Dacochile microsoma* Poinar & Brown: Tanyderidae or Bruchomyiinae? Zootaxa 1162: 19–31.

48. Oberprieler, R. G. 2004. Antliarhininae Schoenherr, 1823 (Coleoptera, Curculionoidea). pp. 829–853 in Brentidae of the World (Coleoptera, Curculionoidea), Sforzi, A. & Bartolozzi, L. (eds.). Regione Piemonte, Torino.

49. Heie, O. E. & Pike, E. M. 1992. New aphids in Cretaceous amber from Alberta (Insecta, Homoptera). Canadian Entomologist 124: 1027–1053.

50. Poinar, Jr., G. O., Gorochov, A. V. & Buckley, R. 2007. *Longioculus burmensis*, n. gen., n. sp. (Orthoptera: Elcanidae) in Burmese

amber. Proceedings of the Entomological Society of Washington 109: 649–655.

51. Borkent, A. 1995. Biting Midges in the Cretaceous Amber of North America (Diptera: Ceratopogonidae). Backhuys Publishers, Leiden, 237 pp.

52. Bakker, R. T. 1986. The Dinosaur Heresies. Kensington Publication Corporation, New York, 481 pp.

53. Strauss, S. Y. & Zangerl, A. R. 2002. Plant-insect interactions in terrestrial ecosystems. pp. 77–106 in Plant-Animal Interactions, Herrera, C. M. & Pellmyr, O. (eds.). Blackwell Science, Oxford.

54. Peters, R. H. 1983. The Ecological Implications of Body Size. Cambridge University Press, Cambridge, 329 pp.

55. Labandeira, C. C. 2002. The history of associations between plants and animals. pp. 26–74 in Plant-Animal Interactions, Herrera, C. M. & Pellmyr, O. (eds.). Blackwell Science, Oxford.

56. Olesen, J. M. & Valido, A. 2003. Lizards as pollinators and seed dispersers: An island phenomenon. Trends in Ecology and Evolution 18: 177–181.

57. Chamberlain, C. J. 1965. The Living Cycads. Hafner, New York, 172 pp.

58. Cooper, M. R. & Goode, D. 2004. The Cycads and Cycad Moths of KwaZulu-Natal. Peroniceras Press, New Germany, KwaZulu-Natal, South Africa, 98 pp.

59. Fullaway, D. T. & Krauss, N. L. H. 1945. Common Insects of Hawaii. Tongg Pub. Co., Honolulu, 228 pp.

60. Hedrick, U. P. 1972. Sturtevant's Edible Plants of the World. Dover Publications, New York, 686 pp.

61. Marvaldi, A. E. 2005. Larval morphology and biology of orycornynine weevils (Belidae). Zoologica Scripta 34: 37–48.

62. Naumann, I. D. 1991. Hymenoptera. pp. 916–1000 in The Insects of Australia, Naumann, I. D. (ed.), Vol. 2, 2nd ed.Comstock Publishing, Ithaca, NY.

63. Miller, D. 1984. Common insects of New Zealand. A. H. & A. W. Reed, Wellington, New Zealand, 179 pp.

64. Tillyard, R. J. 1926. The Insects of Australia and New Zealand. Angus & Robertson, Sydney, Australia, 560 pp.

65. Furniss, R. L. & Carolin, V. M. 1977. Western Forest Insects. Miscellaneous publication No. 1339, United States Department of Agriculture, Forest Service, Washington, D.C. 654 pp.

66. Smith, R. L. 1996. Ecology and Field Biology, 5th ed. Harper Collins, New York, 804 pp.

67. McClure, M. S. 1991. Density-dependent feedback and population cycles in *Adelges tsugae* (Homoptera: Adelgidae) on *Tsuga canadensis*. Environmental Entomology 20: 258–264.

68. Flanders, S. E. 1962. Did the caterpillar exterminate the giant reptile? Journal of Research on the Lepidoptera 1: 85–88.

69. Barrett, P. M. & Willis, K. J. 2001. Did dinosaurs invent flowers? Dinosaur-angiosperm coevolution revisited. Biological Review 76: 411–447.

70. Krassilov, V. A. 2003. Terrestrial Palaeoecology and Global Change. Pensoft Pub., Sofia, Bulgaria, 464 pp.

71. Schoonhoven, L. M., van Loon, J. J. A. & Dicke, M. 2005. Insect-Plant Biology, 2nd ed. Oxford University Press, Oxford, 421 pp.

72. Stevenson, D. W., Norstog, K. J. & Fawcett, P. K. S. 1998. Pollination biology of cycads. pp. 277–294 in Reproductive Biology, Owens, S. J. & Rudall, P. J. (eds.). Royal Botanic Gardens, Kew, England.

73. Krantz, G. W. & Poinar, Jr., G. O., 2004. Mites, nematodes and the multimillion dollar weevil. Journal of Natural History 38:135–141.

74. Krombein, K. V., Norden, B. B., Rickson, M. M. & Rickson, F. R. 1999. Biodiversity of the domatia occupants (ants, wasps, bees, and others) of the Sri Lankan myrmecophyte *Humboldtia laurifolia* Vahl (Fabaceae). Smithsonian Contributions to Zoology 603: 1–34.

75. Danforth, B. N., Sipes, S., Fang, J., & Brady, S.G. 2006. The history of early bee diversification based on five genes plus morphology. Proceedings of the National Academy of Sciences 103:15118–15123.

76. Stephen, W. P., Bohart, G. E. & Torchio, P. F. 1969. The Biology and External Morphology of Bees. Agricultural Experiment Station, Oregon State University, Corvallis, 140 pp.

77. Westrich, P. 1996. Habitat requirements of central European bees and the problems of partial habitats. pp. 1–16 in The Conservation of Bees, Matheson, A., Buchmann, S. L., O'Tolle, C., Westrich, P. & Williams, I. H. (eds.). Academic Press, London.

78. Poinar, Jr., G. O. 2006. Retracing the long journey of the insects. American Scientist 94: 376–378.

79. Free, J. B. 1993. Insect Pollination of Crops. Academic Press, London, 684 pp.

80. Mecke, R., Galileo, M. H. M. & Engels, W. 2001. New records of insects associated with Araucaria trees: Phytophagous Coleoptera and Hymenoptera and their natural enemies. Studies on Neotropical Fauna and Environment 36: 113–124.

81. Harrington, T. C. 1993. Biology and taxonomy of fungi associated with bark beetles. pp. 37–58 in Beetle-Pathogen Interactions in Conifer Forests, Schowalter, T. D. & Filip, G. M. (eds.). Academic Press, London.

82. Farr, D. F., Bills, G. F., Chamuris, G. P. & Rossman, A. Y. 1989. Fungi on Plant and Plant Products in the United States. American Phytopathological Society, St. Paul, MN, 1,252 pp.

83. Sequeira, A. S. & Farrell, B. D. 2001. Evolutionary origins of Gondwanian interaction: How old are Araucaria beetle herbivores? Biological Journal of the Linnean Society 74: 459–474.

84. Holmes, F. W. 1980. Bark beetles, C. ulmi and Dutch Elm Disease. pp. 133–147 in Vectors of Plant Pathogens, Harris, K. F. & Maramorosch, K. (eds.). Academic Press, New York.

85. Russin, J. S., Shain, L., & Nordin, G. L. 1984. Insects as carriers of virulent and cytoplasmic hypovirulent isolates of the chestnut blight fungus. Journal of Economic Entomology 77: 838–846.

86. Harrington, T. C. 2005. Ecology and evolution of mycophagous bark beetles and their fungal partners. pp. 275–291 in Insect-Fungal Associations: Ecology and Evolution. Vega, F. E. & Blackwell, M. (eds.). Oxford University Press, Oxford.

87. Bedding, R. A. 1993. Biological Control of Sirex noctilio using the nematode Deladenus siricidicola. pp. 11–20 in Nematodes and the Biological Control of Insect Pests, Bedding, R., Akhurst, R. & Kaya, H. (eds.). CSIRO Publication, East Melbourne, Australia.

88. Mamiya, Y. 1972. Pine wood nematode, Bursaphelenchus lignicolus Mamiya & Kiyohara, as a causal agent of pine wilting disease. Review of Plant Protection Research 5: 46–60.

89. Agrios, G. N. 1987. Plant Pathology, 2nd ed., Academic Press, New York, 703 pp.

90. Ziller, W. G. 1974. The Tree Rusts of Western Canada. Canadian Forestry Service Publication 1329, 272 pp.

91. Poinar, Jr., G. O. & Brown, A. E. 2003. A non-gilled hymenomycete in Cretaceous amber. Mycological Research 107: 763–768.

92. Alexopoulos, A. 1952. Introductory Mycology. John Wiley & Sons, New York, 483 pp.

93. Bessey, E. A. 1950. Morphology and Taxonomy of Fungi. The Blackiston Co., Philadelphia, 791 pp.

94. Newton, Jr., A. F. 1984. Mycophagy in Staphylinoidea (Coleoptera). pp. 302–353 in Fungus-Insect Relationsips, Wheeler, Q. & Blackwell, M. (eds.). Columbia University Press, New York.

95. Harris, K. F. & Maramorosch, K. 1980. Vectors of Plant Pathogens. Academic Press, New York, 467 pp.

96. Teakle, D. S. & Pares, R. D. 1977. Potyvirus (Potato Virus Y) Group, pp. 311–325 in The Atlas of Insect and Plant Viruses, Maramorosch, K. (ed.). Academic Press, New York.

97. Morales, F. J., Lozano, I., Sedano, R., Castaño, M. & Arroyave, J. 2002. Partial characterization of a potyvirus infecting African oil palm in South America. Journal of Phytopathology 150: 297–301.

98. Koteja, J. 1989. *Inka minuta* gen. et sp. n. (Homoptera, Coccinea) from Upper Cretaceous Taymyrian amber. Annales Zoologici 43: 77–101.

99. Koteja, J. 2004. Scale insects (Hemiptera: Coccinea) from Cretaceous Myanmar (Burmese) amber. Journal of Systematic Palaeontology 2: 109–114.

100. Baker, W. L. 1972. Eastern Forest Insects. United States Department of Agriculture, Forest Service, Miscellaneous Publication No. 1175, Washington, D.C., 642 pp.

101. Muniyappa, V. 1980. Whiteflies. pp. 39–85 in Vectors of Plant Pathogens, Harris, K. F. & Maramorosch, K. (eds.). Academic Press, New York, 467 pp.

102. Thresh, J. M. 1991. The ecology of tropical plant viruses. Plant Pathology 40: 324–339.

103. Manauté, J., Jaffré, T., Veillon, J.-M. & Kranitz, M.-L. 2003. Revue des Araucariaceae de Nouvelle-Caledonie, IRD/Province Sud, Nouméa, New Caledonia, 28 pp.

104. Rasnitsyn, A. P. 1992. *Strashila incredibilis*, a new enigmatic mecopteroid insect with possible siphonapteran affinities from the Upper Jurassic of Siberia. Psyche 99: 323–333.

105. Laurence, B. R. 1954. The larval inhabitants of cow pats. The Journal of Animal Ecology 23: 234–260.

106. Hanski, I. & Cambefort, Y. 1991. Dung Beetle Ecology. Princeton University Press, Princeton, NJ, 481 pp.

107. Nikolayev, G. V. 1993. The taxonomic placement in the subfamily Aphodiinae (Coleoptera, Scarabaeidae) of the new genus of

Lower Cretaceous scarabid beetles from Transbaykal. Paleontological Journal 27: 1–8.

**108.** Krell, F. T. 2000. The fossil record of Mesozoic and Tertiary Scarabaeoidea (Coleoptera: Polyphaga). Invertebrate Taxonomy 14: 871–905.

**109.** Buss, I. O. 1990. Elephant Life. Iowa State University Press, Ames, 151 pp.

**110.** Lewin, R. A. 1999. Merde. Random House, New York, 187 pp.

**111.** Halffter, G. 1972. Eléments anciens de l'entomofauna néotropicale: Ses implications biogéographiques. In Biogéographie et Liaisons Intercontinentales au Cours du Mésozoique. 17th International Congress of Zoology, Monte Carlo 1: 1–40.

**112.** Jeannel, R. 1942. La Genèse des Faunes Terrestres. Press Univ. de France, Paris, 513 pp.

**113.** Benton, M. J. 1984. The Dinosaur Encyclopedia. Simon & Schuster, New York, 188 pp.

**114.** Lambert, D. 1990. The Dinosaur Data Book. Avon Books, New York, 320 pp.

**115.** Chin, K., Tokaryk, T. T., Erickson, G. M. & Calk, L. C. 1998. A king-sized theropod coprolite. Nature 393: 680–682.

**116.** Chin, K. & Gill, B. D. 1996. Dinosaurs, dung beetles and conifers, participants in a Cretaceous food web. Palaios 11: 280–285.

**117.** Young, O. P. 1981. The attraction of Neotropical Scarabaeinae (Coleoptera: Scarabaeidae) to reptile and amphibian fecal material. Coleopterists Bulletin 35: 345–348.

**118.** Halffter, G. & Matthews, E. 1966. The natural history of dung beetles of the subfamily Scarabaeinae (Coleoptera, Scarabaeidae). Folia Entomologica Mexicana 12–14: 1–312.

**119.** Petrovitz, R. 1962. Neue und interessante Scarabaeidae aus dem vorderen Orient. Reichenbachia 1: 107–124.

**120.** Hubbard, H. G. 1894. The insect guests of the Florida land tortoise. Insect Life 6: 302–315.

**121.** Young, F. N. & Goff, C. C. 1939. An annotated list of the arthropods found in the burrows of the Florida gopher tortoise, *Gopherus polyphemus* (Daudin). The Florida Entomologist 22: 53–62.

**122.** Brach, V. 1977. Larvae of *Onthophagus p. polyphemi* Hubbard and *Onthophagus tuberculifrons* Harold (Coleoptera: Scarabaeidae).

Bulletin of the Southern California Academy of Sciences 76: 66–68.

123. Gill, B. D. 1991. Dung beetles in tropical American forests. pp. 211–229 in Dung Beetle Ecology, Hanski, I. & Cambefort, Y. (eds.). Princeton University Press, Princeton, New Jersey.

124. Catts, E. P. & Haskell, N. H. 1990. Entomology and Death: A Procedural Guide. Joyce's Print Shop, Clemson, 182 pp.

125. Fretey, J. & Babin, R. 1998. Arthropod succession in leatherback turtle carrion and implications for determination of the post-mortem interval. Marine Turtle Newsletter 79: 4–7.

126. McAlpine, J. F. 1970. First record of calypterate flies in the Mesozoic Era. The Canadian Entomologist 102: 342–346.

127. Hasiotis, S. T. & Fiorillo, A. R. 1997. Dermestid beetle borings in dinosaur bones, Dinosaur National Monument, Utah: Additional keys to bone bed taphonomy. Abstracts of the 1997 31st annual South-Central and 50th Annual Rocky Mountain sections of the Geological Society of America Meetings. No. 14609: 13.

128. Laws, R. R., Hasiotis, S. T., Fiorillo, A. R., Chure, D. J., Breithaupt, B. H. & Horner, J. R. 1996. The demise of a Jurassic Morrison Dinosaur after death. Three cheers for the dermestid beetle. Abstracts of the 1996 Annual Meeting of the Geological Society of America, A-299.

129. Rogers, R. R. 1992. Non-marine borings in dinosaur bones from the Upper Cretaceous Two Medicine Formation, Northwestern Montana. Journal of Vertebrate Paleontology 12: 528–531.

130. Hinton, H. E. 1945. A Monograph of the Beetles Associated with Stored Products. Printed by order of the trustees of the British Museum, London, 443 pp.

131. Cockerell, T. D. A. 1917. Arthropods in Burmese amber. Psyche 24 : 40–42.

132. Price, P. W. 1997. Insect Ecology, 3rd ed. Wiley, New York, 874 pp.

133. Janzen, D. H. 1983. Costa Rican Natural History. The Univerisity of Chicago Press, Chicago, 816 pp.

134. DeFoliart, G. 1992. Insects as human food. Crop Protection 11: 395–399.

135. Poinar, Jr., G. & Boucot, A. J. 2006. Evidence of intestinal parasites of dinosaurs. Parasitology 133: 245–249.

136. Morris, G. K. & Gwynne, D. T. 1978. Geographical distribution

and biological observations of *Cyphoderris* (Orthoptera: Haglidae) with a description of a new species. Psyche 85: 147–167.

**137.** Bodenheimer, F. S. 1951. Insects as human food. Dr. W. Junk, the Hague, 352 pp.

**138.** Menzel, P. & D'Aluisio, F. 1998. Man Eating Bugs, the Art and Science of Eating Insects. Ten Speed Press, Berkeley, CA, 192 pp.

**139.** Isaacs, J. 1987. Bush Food. Aboriginal Food and Herbal Medicine. Weldon Publishers, Willoughby, New South Wales, 256 pp.

**140.** Thomson, R. C. M. 1951. Mosquito Behavior in Relation to Malaria Transmission and Control in the Tropics. Edward Arnold and Co., London, 219 pp.

**141.** Downs, J. A. 1970. The ecology of blood-sucking Diptera: An evolutionary perspective. pp. 232–258, in Ecology and Physiology of Parasites, Fallis, A. M. (ed.). University of Toronto Press, Toronto.

**142.** Lehane, M. J. 1991. Biology of Blood-Sucking Insects. Harper Collins Academic, London, 288 pp.

**143.** Schlein, Y. & Warburg, A. 1986. Phytophagy and the feeding cycle of *Phlebotomus papatasi* (Diptera: Psychodidae) under experimental conditions. Journal of Medical Entomology 23: 11–15.

**144.** Chiasson, R. B., Bentley, D. L. & Lowe, C. H. 1989. Scale morphology in *Agkistrodon* and closely related crotaline genera. Herpetology 43: 430–438.

**145.** Lull, R, S. & Wright, N. E. 1942. Hadrosaurian dinosaurs of North America. Geological Society of America Special Papers 40: 1–242.

**146.** Osborn, H. F. 1912. Integument of the iguanodont dinosaur *Trachodon*. Memoirs of the American Museum of Natural History 1: 33–54.

**147.** Brown, B. 1916. *Corythosaurus casuarius:* Skeleton, musculature and epidermis. Bulletin of the American Museum of Natural History 35: 709–716.

**148.** Gilmore, C. W. 1946. Notes on recently mounted reptile fossil skeletons in the United States National Museum. Proceedings of the United States National Museum 96: 195–203.

**149.** Sternberg, C. M. 1925. Integument of *Chasmosaurus belli*. The Canadian Field-Naturalist 39:108–110.

**150.** Hooley, R. W. 1917. On the integument of *Iguanodon bernissartensis* Boulenger, and of *Morosaurus becklesii* Mantell. Geological Magazine 4: 148–150.

151. Czerkas, S. 1994. The history and interpretation of sauropod skin impressions. pp. 173–182 in Aspects of Sauropod Paleobiology, Lockley, M. G., dos Santos, V. F., Meyer, C. A. & Hunt, A. P. (eds.). Gaia 10: 1–279.

152. Currie, P. J. 2000. Feathered dinosaurs. pp. 183–189 in The Scientific American Book of Dinosaurs, Paul, G. S. (ed.). St. Martin's Griffin, New York.

153. Markle, W. H. & Makhoul, K. 2004. Cutaneous leishmaniasis: Recognition and treatment. American Family Physician 69:1455–1460.

154. Smith, H. R. 1946. Handbook of Lizards. Comstock Publishing, Ithaca, NY, 557 pp.

155. Reid, R. E. H. 1997. Dinosaurian physiology: The case for "intermediate dinosaurs." pp. 449–473 in The Complete Dinosaur, Farlow, J. D. & Brett-Surman, M. K. (eds.). Indiana University Press, Bloomington.

156. Borkent, A. 2000. Biting midges (Ceratopogonidae: Diptera) from Lower Cretaceous Lebanese amber with a discussion of the diversity and patterns found in other ambers. pp. 355–451 in Studies on Fossils in Amber, with Particular Reference to the Cretaceous of New Jersey, Grimaldi, D. (ed.). Backhuys Publishers, Leiden.

157. Mullen, G. R. 2002. Biting midges (Ceratopogonidae). pp. 163–183 in Medical and Veterinary Entomology, Mullen, G. and Durden, L. (eds.). Academic Press, San Diego, CA.

158. Meiswinkel, R. 1992. Afrotropical *Culicoides: C. (Avaritia) laxodontis* sp. nov., a new member of the *imicola* group (Diptera: Ceratopogonidae) associated with the African elephant in the Kruger National Park, South Africa. Onderstepoort Journal of Veterinary Research 59: 145–160.

159. Seppa, N. 2004. Soldiers in Iraq coming down with parasitic disease. Science News 166: 53.

160. Mullens, B. A., Barrows, C. & Borkent, A. 1997. Lizard feeding by *Leptoconops (Brachyconops) californiensis* (Diptera: Ceratopogonidae) on desert sand dunes. Journal of Medical Entomology 34: 735–737.

161. Auezova, G. 1998. The biting midges (Diptera: Ceratopogonidae)—bloodsuckers of reptiles and bats—as possible collabora-

tors of arboviruses circulating in nature. Parasitology International 47 (Suppl.): 306.

162. Wirth, W. W. & Hubert, A. A. 1962. The species of *Culicoides* related to *piliferus* Root and Hoffman in eastern North America (Diptera, Ceratopogonidae). Annals of the Entomological Society of America 55: 182–195.

163. Borkent, A. 1995. Biting midges (Ceratopogonidae: Diptera) feeding on a leatherback turtle in Costa Rica. Brenesia 43–44: 25–30.

164. Szadziewski, R. & Poinar, Jr., G. O. 2005. Additional biting midges (Diptera: Ceratopogonidae) from Burmese amber. Polska Pismo Entomologiczne 74: 349–362.

165. Purse, B. V., Mellor, P. S., Rogers, D. J., Samuel, A. R., Mertens, P. P. C. & Baylis, M. 2005. Climate change and the recent emergence of bluetongue in Europe. Nature Reviews Microbiology 3: 171–181.

166. Wirth, W. W. & Lee, D. J. 1958. Australasian Ceratopogonidae (Diptera: Nematocera). Part VIII: A new genus from Western Australia attacking man. Proceedings of the Linnean Society of New South Wales 83: 337–339.

167. Poinar, Jr., G. & Telford, Jr., S. R. 2005. *Paleohaemoproteus burmacis* gen. n., sp. n., (Haemospororida: Plasmodiidae) from an Early Cretaceous biting midge (Diptera: Ceratopogonidae). Parasitology 131: 1–6.

168. Atkinson, C. T. & Van Riper III, C. 1991. Pathogenicity and epizootiology of avian Haematozoa: *Plasmodium, Leucocytoozoon*, and *Haemoproteus*. pp. 19–48 in Loye, J. E. & Zuk, M. (eds.). Bird-Parasite Interactions: Ecology, Evolution, and Behaviour. Oxford University Press, New York.

169. Fallis, A. M. & Bennett, G. F. 1961. Ceratopogonidae as intermediated hosts for *Haemoproteus* and other parasites. Mosquito News 21: 21–28.

170. Paperna, I. & Landau, I. 1991. *Haemoproteus* (Haemosporidia) of lizards. Bulletin du Museum National Histoire, Paris 13: 309–349.

171. Doherty, R. L., Carley, J. G., Standfast, H. A., Dyce, A. L., Kay, B. H. & Snowdon, W. A. 1973. Isolation of arboviruses from mosquitoes, biting midges, sandflies and vertebrates collected in Queens-

land, 1969 and 1970. Transactions of the Royal Society of Tropical Medicine and Hygiene 67: 536–543.

172. Poinar, Jr., G. O. & Poinar, R. 2005. Fossil evidence of insect pathogens. Journal of Invertebrate Pathology 89: 243–250.

173. Anderson, R. C. 2000. Nematode Parasites of Vertebrates, 2nd ed. CABI Publishing, Wallingford, England, 650 pp.

174. Azar, D., Nel, A., Solignac, M., Paicheler, J.-C. & Bouchet, F. 1999. New genera and species of psychodoid flies from the Lower Cretaceous amber of Lebanon. Palaeontology 42: 1101–1136.

175. Rutledge, L. C. & Gupta, R. K. 2002. Moth flies and sand flies (Psychodidae). pp. 147–161 in Medical and Veterinary Entomology, Mullen, G. & Durden, L. (eds.). Academic Press, San Diego, CA.

176. Chaniotis, B. N. 1967. The biology of California *Phlebotomus* (Diptera: Psychodidae) under laboratory conditions. Journal of Medical Entomology 4: 221–233.

177. Klein, T. A., Young, D. G., Telford, Jr., S. R. & Kimsey, R. 1987. Experimental transmission of *Plasmodium mexicanum* by bites of infected *Lutzomyia vexator* (Diptera: Psychodidae). Journal of the American Mosquito Control Association 3: 154–164.

178. Da Silva, O. S. & Grunewald, J. 2000. First detection of blood in the gut of wild caught male sandflies in southern Brazil. Studies on Neotropical Fauna and Environment 35: 201–202.

179. Perfilev, P. P. 1968. Phlebotomidae (sandflies). In Fauna of U.S.S.R. (Diptera), Vol. 33, No. 2, Akademiya Nauk SSSR, Moscow. (Translated from Russian by the Israel program for scientific translations, Jerusalem.) 363 pp.

180. Boucot, A. J., Chen, Xu & Scotese, C. R. 2008. Preliminary compilation of Cambrian through Miocene climatically sensitive deposits. Memoirs of the Society of Economic Paleontologists and Mineralogists, Norman, OK (in press).

181. Poinar, Jr., G. O., Jacobson, R. L. & Eisenberger, C. L. 2006. Early Cretaceous phlebotomine sand fly larvae (Diptera: Psychodidae). Proceedings of the Entomological Society of Washington 108: 785–792.

182. Thomson, L. A. J. 2006. *Agathis macrophylla* (Pacific kauri). pp. 29–40 in Traditional Trees of Pacific Islands, Elevitch, C. R. (ed.). Holualoa, Hawaii.

183. Telford, Jr., S. R. 1995. The kinetoplastid hemoflagellates of reptiles. pp. 161–223 in Parasitic Protozoa, 2nd ed., Kreier, J. P. (ed.). Vol. 10. Academic Press, San Diego.

184. Brygoo, E. R. 1963. Hematozoaires de Reptiles malgaches. 1. *Trypanosoma therezieni* n. sp. parasite des chameleons de Madagascar. Infestation naturelle et experimentale. Archives de Institute de Pasteur Madagascar 31: 133–141.

185. Dedet, J.-P. 2002. Current status of epidemiology of leishmaniases. pp. 1–10 in Leishmania, Farrell, J. P.(ed.). Kluver Academic Publishers, Dordrecht.

186. Montoya-Lerma, J., Cadena, H., Oviedo, M., Ready, P. D., Barazarte, R., Travi, B. L. & Lane, R. P. 2002. Comparative vectorial efficiency of *Lutzomyia evansi* and *Lu. longipalpis* for transmitting *Leishmanial chagasi*. Acta Tropica 85: 19–29.

187. Yuval, B. 1991. Populations of *Phlebotomus papatasi* (Diptera: Psychodidae) and the risk of *Leishmanial major* transmission in three Jordan Valley habitats. Journal of Medical Entomology 28: 492–495.

188. Schlein, Y. Warburg, A., Schnur, L. F. & Gunders, A. E. 1982. Leishmaniasis in the Jordan Valley II. Sandflies and transmission in the central endemic area. Transactions of the Royal Society of Tropical Medicine and Hygiene 76: 582–589.

189. Ayala, S. C. 1970. Lizard malaria in California; description of a strain of *Plasmodium mexicanum*, and biogeography of lizard malaria in Western North America. The Journal of Parasitology 56: 417–425.

190. Poinar, Jr., G. O. & Poinar, R. 1999. The Amber Forest, Princeton University Press, Princeton, NJ, 239 pp.

191. Poinar, Jr., G. O., Zavortink, T. J., Pike, T. & Johnston, P. A. 2000. *Paleoculicis minutus* (Diptera: Culicidae) n. gen., n. sp., from Cretaceous Canadian amber, with a summary of described fossil mosquitoes. Acta Geologica Hispanica 35: 119–128.

192. Day, J. F. & Curtis, G. A. 1983. Opportunistic blood-feeding on egg-laying sea turtles by salt marsh mosquitoes (Diptera: Culicidae). Florida Entomologist 66: 359–360.

193. Fretey, J. 1989. Attaques diurnes ou nocturnes de *Tortues luths* par des Tabanidés et autres Diptères hématophages en Guyane francaise et au Surinam. L'Entomologiste 45: 237–244.

**194.** Foster, W. A. & Walker, E. D. 2002. Mosquitoes (Culicidae).pp. 203–262 in Medical and Veterinary Entomology, Mullen, G. & Durden, L. (ed.). Academic Press, San Diego, CA.

**195.** Horsefall, W. R. 1955. Mosquitoes, Their Bionomics and Relation to Disease. The Ronald Press, New York, 723 pp.

**196.** Klein, T. A., Young, D. G. & Telford, Jr., S. R. 1987. Vector incrimination and experimental transmission of *Plasmodium floridense* by bites of infected *Culex (Melanoconion) erraticus*. Journal of the American Mosquito Control Association 3: 165–175.

**197.** Jordan, H. B. 1964. Lizard malaria in Georgia. Journal of Protozoology 11: 562–566.

**198.** Telford, S. R. 1994. Plasmodia of Reptiles. pp. 1–71 in Parasitic Protozoa, 2nd ed., Kreier, J. P. (ed.). Vol. 7. Academic Press, San Diego.

**199.** Perkins, S. L. 2001. Phylogeography of Caribbean lizard malaria: Tracing the history of vector-borne parasites. Journal of Evolutionary Biology 14: 34–45.

**200.** Klein, T. A., Akin, D. C., Young, D. G. & Telford, Jr., S. R.1988. Sporogony, development and ultrastructure of *Plasmodium floridense* in *Culex erraticus*. International Journal for Parasitology 18: 711–719.

**201.** Marra, P. P., Griffing, S., Caffrey, A., Kilpatrick, A. M., McLean, R., Brand, C., Saito, E., Dupris, A. P., Kramer, L. & Novak, R. 2004. West Nile virus and wildlife. BioScience 54: 393–402.

**202.** Miller, D., Manual, M., Baldwin, C., Burtle, C., Ingram, G., Hines, M. & Frazier, K. 2003. West Nile virus in farmed alligators. Emerging Infectious Diseases 9: 641–646.

**203.** Frank, W. 1981. Endoparasites. pp. 291–358 in Diseases of the Reptilia, vol. 1, Cooper, J. E. & Jackson, O. F. (eds.). Academic Press, London.

**204.** Adler, P. H. & McCreadie, J. W. 2002. Black flies (Simuliidae). pp. 185–202 in Medical and Veterinary Entomology, Mullen, G. & Durden, L. (eds.). Academic Press, San Diego, CA.

**205.** Crosskey, R. W. 1990. The Natural History of Blackflies. John Wiley and Sons, New York, 711 pp.

**206.** Jell, P. A & Duncan, P. M. 1986. Invertebrates, mainly insects, from the freshwater, Lower Cretaceous, Koonwarra fossil bed (Korumburra Group), South Gippsland, Victoria. pp. 111–205 in Plants and Invertebrates from the Lower Cretaceous Koonwarra

Fossil Bed, South Gippsland, Victoria, Jell, P. A & Roberts, J. (eds.). Association of Australasian Palaeontologists, Sydney.

207. Kalugina. N. S. 1991. New Mesozoic Simuliidae and Lepto-cononopidae and the origin of bloodsucking in the lower Dipteran insects. Paleontological Journal 1: 69–80.

208. Currie, D. C. & Grimaldi, D. 2000. A new black fly (Diptera: Simuliidae) genus from mid Cretaceous (Turonian) amber of New Jersey. pp. 473–485 in Studies on Fossils in Amber, with Particular Reference to the Cretaceous of New Jersey, Grimaldi, D. (ed.). Backhuys Publishers, Leiden.

209. Smith, C. D. 1969. The effects of temperature on certain life stages of Simuliidae (Diptera). M.S. thesis. University of Durham, 122 pp.

210. Martins-Neto, R. G. & Kucera-Santos, J. C. 1994. Um nôvo gênero e nôva espécie de mutuca (Insecta, Diptera, Tabanidae) da Formação Santana (Cretáceo Inferior), Bacia do Araripe, Nordeste do Brasil. Acta Geologica Leopoldensia 17: 289–297.

211. Coram, R., Jarzembowski, E. A. & Ross, A. J. 1995. New record of Purbeck fossil insects. Proceedings of the Dorset Natural Historical and Archaeological Society 116: 145–150.

212. Ren, D. 1998. Late Jurassic Brachycera from Northeastern China (Insects: Diptera). Acta Zootaxonomica Sinica 23: 65–83.

213. Mostovski, M. B., Jarzembowski, E. A. & Coram, R. A. 2003. Tabanids and athericids (Diptera: Tabanidae, Athericidae) from the Lower Cretaceous of England and Transbaikalia. Paleontological Zhurnal 2: 57–64.

214. Ferreira, R. L. M., Henriques, A. L. & Rafael, J. A. 2002. Activity of tabanids (Insects: Diptera: Tabanidae) attacking the reptiles *Caiman crocodilus* (Linn.) (Alligatoridae) and *Eunectes murinus* (Linn.) (Boidae), in the Central Amazon, Brazil. Memórias do Instituto Oswaldo Cruz 97:133–136.

215. Philip, C. B. 1983. A unique, divergent developmental dependence of a Galapagos tabanid (Diptera, Tabanidae). Wasmann Journal of Biology 41: 47–49.

216. Fretey, J. 1989. Attaques diurnes ou nocturnes de *Tortues luths* par des Tabanidés et autres Diptères hématophages en Guyane francaise et au Surinam. L'Entomologiste 45: 237–244.

217. Sterling, C. R. & de Guisti, D. L. 1974. Fine structure of differentiating oocysts and mature sporozoites of *Haemoproteus*

*metchnikovi* in its intermediate host *Chrysops callidus*. Journal of Protozoology 21: 276–283.

218. Mullens, B. A. 2002. Horse flies and deer flies (Tabanidae). pp. 263–277 in Medical and Veterinary Entomology, Mullen, G. & Durden, L. (eds.). Academic Press, San Diego, CA.

219. Fox, I., Fox, R. I. & Bayona, I. G. 1966. Fleas feed on lizards in the laboratory in Puerto Rico. Journal of Medical Entomology 2: 395–396.

220. Durden, L. A. & Traub, R. 2002. Fleas (Siphonaptera). pp. 103–125 in Medical and Veterinary Entomology, Mullen, G. & Durden, L. (eds.). Academic Press, San Diego, CA.

221. Ponomarenko, A. G. 1976. A new insect from the Cretaceous of Transbaikalia, a possible parasite of pterosaurians. Paleontological Journal 3: 339–343.

222. Ponomarenko, A. G. 1986. Insects in the early Cretaceous ecosystems of the West Mongolia. Transactions of the Joint Soviet-Mongolian Palaeontological Expedition 28: 1–214.

223. Durden, L. A. 2002. Lice (Phthiraptera). pp. 45–65 in Medical and Veterinary Entomology, Mullen, G. & Durden, L. (eds.). Academic Press, San Diego, CA.

224. Rasnitsyn, A. P. & Zherikhin, V. V. 1999. First fossil chewing louse from the Lower Cretaceous of Baissa, Transbaikalia (Insects, Pediculida = Phthiriaptera, Saurodectidae fam. n.). Russian Entomology Journal 8: 253–255.

225. Derylo, A. 1970. Mallophaga as a reservoir of *Pasteurella multocida*. Acta Parasitologica Polonica 17: 301–313.

226. Klompen, J. S. & Grimaldi, D. 2001. First Mesozoic record of a parasitiform mite: A larval argasid tick in Cretaceous amber (Acari: Ixodida: Argasidae). Annals of the Entomological Society of America 94: 10–15.

227. Keirans, J. E. & Gattis, G. I. 1986. *Amblyomma arianae*, n. sp. (Acari: Ixodidae), a parasite of *Alsophis portoricensis* (Reptilia: Colubridae) in Puerto Rico. Journal of Medical Entomology 23: 622–625.

228. Frey, F. L. 1991. Reptile Care. An Atlas of Diseases and Treatments, Vol. 1. T. F. H. Publications, Neptune City, NJ, 341 pp.

229. Arthur, D. R. 1961. Ticks and Disease. Row, Peterson & Co., Evanston, IL, 445 pp.

230. Sonenshine, D. E., Lane, R. S. & Nicholson, W. L. 2002. Ticks

(Ixodida). pp. 517–558 in Medical and Veterinary Entomology, Mullen, G., & Durden, L. (eds.). Academic Press, San Diego, CA.

231. de la Fuente, J. 2003. The fossil record and the origin of ticks (Acari: Parasitiformes: Ixodida). Experimental and Applied Acarology 29: 331–344.

232. Parola. P & Raoult, D. 2001. Tick-borne bacterial diseases emerging in Europe. Clinical and Microbiological Infections 7: 80–83.

233. Lane, R. S. & Quistad, G. B. 1998. Borrelicidal factor in the blood of the western fence lizard (*Sceloporus occidentalis*). Journal of Parasitology 84: 29–34.

234. Cooper, J. E. 1981. Bacteria. pp. 165–191 in Diseases of the Reptilia, Vol. 1, Cooper, J. E. & Jackson, O. F. (eds.). Academic Press, New York.

235. Hoff, G. L. 1984. Q fever. pp. 101–106 in Diseases of Amphibians and Reptiles, Hoff, G. L., Frye, F. L. & Jacobson, E. R. (eds.). Plenum Press, New York.

236. Mullen, G. R. & O'Conner, B. M. 2002. Mites (Acari). pp. 449–516 in Medical and Veterinary Entomology. Mullen, G. & Durden, L. (eds.). Academic Press, San Diego, CA.

237. Lawrence, R. F. 1935. The prostigmatic mites of South African lizards. Parasitology 27: 1–45.

238. Lawrence, R. F. 1936. The prostigmatic mites of South African lizards. Parasitology 28: 1–39.

239. Goodwin, Jr., M. H. 1954. Observations on the biology of the lizard mite *Geckobiella texana* (Banks) 1904 (Acarina: Pterygosomidae). Parasitology 40: 54–59.

240. Davidson, J. A. 1958. A new species of lizard mite and a generic key to the family Pterygosomidae. Proceedings of the Entomological Society of Washington 60: 75–79.

241. Lawrence, R. F. 1949. The larval Trombiculid mites of South African vertebrates. Annals of the Natal Museum 11: 405–486.

242. Moravec, F. 2001. Trichinelloid Nematodes Parasitic in Cold-Blooded Vertebrates. Academia, Prague, 429 pp.

243. Yamaguti, S. 1961. Systema Helmithum, Vol.3, Parts 1 and 2 of The Nematodes of Vertebrates. Interscience Publishers, New York, 1,261 pp.

244. Fincher, G. T., Stewart, T. B. & Davis, R. 1969. Beetle intermedi-

ate hosts for swine spirurids in southern Georgia. Journal of Parasitology 55: 355–358.

245. Poinar, G. O. & Vaucher, C. 1972. Cycle larvaire de *Physaloptera retusa* Rudolphi, 1819 (Nematoda, Physalopteridae), parasite d'un Lézard sud-amèricain. Bulletin du Musèum National d'Histoire Naturelle 74: 1321–1327.

246. Goldberg, S. R., Bursey, C. R. & Morando, M. 2004. Metazoan endoparasites of 12 species of lizards from Argentina. Comparative Parasitology 71: 208–214.

247. Sprent, J. F. A. 1978. Ascaridoid nematodes of amphibians and reptiles: *Polydelphis, Travassosascaris* n.g. and *Hexametra*. Journal of Helminthology 52: 355–384.

248. Sprent, J. F. A. 1984. Ascaridoid nematodes. pp. 219–245 in Diseases of Amphibians and Reptiles, Hoff, G. L., Frye, F. L. & Jacobson, E. R. (eds.). Plenum Press, New York.

249. Poinar, Jr., G. O., Chabaud, A. G. & Bain, O. 1989. *Rabbium paradoxus* sp. n. (Seuratidae: Skrjabinelaziinae) maturing in *Camponotus castaneus* (Hymenoptera: Formicidae). Proceedings of the Helminthological Society of Washington 56: 120–124.

250. Brooks, D. R. 1984. Platyhelminths. pp. 247–258 in Diseases of Amphibians and Reptiles, Hoff, G. L., Frye, F. L. & Jacobson, E. R. (eds.). Plenum Press, New York.

251. Conn, D. B. 1985. Life cycle and postembryonic development of *Oochoristica anolis* (Cyclophyllidea: Linstowiidae). Journal of Parasitology 71: 10–16.

252. Poinar, Jr., G. O. & Hess, R. 1982. Ultrastructure of 40-million-year-old insect tissue. Science 215: 1241–1242.

253. Poinar, Jr., G. O. & Hess, R. 1985. Preservative qualities of recent and fossil resins: Electron micrograph studies on tissues preserved in Baltic amber. Journal of Baltic Studies 16: 222–230.

254. Poinar, Jr., G. O. & Poinar, R. 2004. *Paleoleishmania proterus* n. gen., n. sp., (Trypanosomatidae: Kinetoplastida) from Cretaceous Burmese amber. Protista 155: 305–310.

255. Poinar, Jr., G. O. 2004. *Palaeomyia burmitis* gen. n., sp. n. (Phlebotomidae: Diptera), a new genus of Cretaceous sand flies with evidence of blood sucking habits. Proceedings of the Entomological Society of Washington 106: 598–605.

256. Altman, P. L. 1961. Blood and Other Body Fluids. Federation of

American Societies for Experimental Biology, Washington, D.C., 540 pp.

257. Baker, J. R. 1976. Biology of the trypanosomes of birds. pp. 131–174 in Biology of the Kinetoplastida, Vol. 1, Lumsden, W. H. R. & Evans, D. A. (eds.). Academic Press, New York.

258. Wilson, V. C. L. & Southgate, B. A. 1979. Lizard Leishmania. pp. 241–268 in Biology of the Kinetoplastida, Vol. 2, Lumsden W. H. R. & Evans D. A. (eds.). Academic Press, New York.

259. Telford, Jr., S. R. 1984. Haemoparasites of Reptiles. pp. 385–517 in Diseases of Amphibians and Reptiles, Hoff, G. L., Frye, F. L. & Jacobson, E. R. (eds.). Plenum Press, New York.

260. Poinar, Jr., G. O. & Poinar, R. 2004. Evidence of vector-borne disease of Early Cretaceous reptiles. Vector-Borne and Zoonotic Diseases 4: 281–284.

261. Poinar, Jr., G. & Telford, Jr., S. R. 2005. *Paleohaemoproteus burmacis* gen. n., sp. n., (Haemospororida: Plasmodiidae) from an Early Cretaceous biting midge (Diptera: Ceratopogonidae). Parasitology 131: 1–6.

262. Anderson, B., Friedman, H. & Bendinelli, M. 2006. Microorganisms and Bioterrorism. Springer, New York, 240 pp.

263. Soto, C. 2006. Prions, the New Biology of Proteins. CRC Press, Boca Raton, FL, 167 pp.

264. Villarreal, L. P. 2005. Viruses and the Evolution of Life. ASM Press, Washington, D.C., 395 pp.

265. Reisen, W. K. 2002. Epidemiology of vector-borne diseases. pp. 15–27 in Medical and Veterinary Entomology, Mullen, G. & Durden, L. (eds.). Academic Press, San Diego, CA.

266. Thompson, P. E. 1944. Changes associated with acquired immunity during initial infections in saurian malaria. Journal of Infectious Diseases 75: 138–149.

267. Tkach, J. R. 1983. Evolutionary immaturity of B-cell function as a possible cause of the Upper Cretaceous extinction of orders Saurischia and Ornithischia. Unpublished document, 24 pp.

268. Moodie, R. L. 1923. Paleopathology:An Introduction to the Study of Ancient Evidences of Disease. University of Illinois Press, Urbana, 567 pp.

269. Schweitzer, M. H., Wittmeyer, J. L., Horner, J. R. & Toporski, J. K. 2005. Soft-tissue vessels and cellular preservation in *Tyrannosaurus rex*. Science 307: 1952–1955.

270. Clausen, H. J. & Duran-Reynals, F. 1937. Studies in the experimental infection of some reptiles, amphibia and fish with *Serratia anolium*. American Journal of Pathology 13 : 441–451.

271. Poinar, Jr., G. O. & Thomas, G. M. 1984. Laboratory Guide to Insect Pathogens and Parasites. Plenum Press, New York, 392 pp.

272. Fauquat, C. M. & Fargette, D. 2005. International Committee on taxonomy of viruses and the 3,142 unassigned species. Virology Journal 2: 64–74.

273. Taylor, T. N. & Taylor, E. L. 1993. The Biology and Evolution of Fossil Plants. Prentice Hall, Englewood Cliffs, NJ, 982 pp.

274. Rossman, A. Y., Farr, D. F., Bills, G. F. & Chamuris, G. P. 1989. Fungi on Plants and Plant Products in the United States. The American Phytopathological Society, St. Paul, MN, 1,252 pp.

275. Migaki, G., Jacobson, E. R. & Casey, H. W. 1984. Fungal diseases in reptiles. pp. 183–204 in Diseases of Amphibians and Reptiles, Hoff, G. L., Frye, F. L. & Jacobson, E. R. (eds.). Plenum Press, New York.

276. Kisla, T. A., Cu-Unjieng, A., Singler, L., & Sugar, J. 2000. Medical management of *Beauveria bassiana* keratitis. Cornea 19: 405–406.

277. Maggenti, A. 1981. General Nematology. Springer-Verlag, New York, 372 pp.

278. Poinar, Jr., G. O. 1975. Entomogenous Nematodes. Brill, Leiden, 317 pp.

279. Poinar, Jr. G. O. 1983. The Natural History of Nematodes. Prentice Hall, Englewood Cliffs, NJ, 323 pp.

280. Garnham, P. C. C. 1966. Malaria Parasites and Other Haemosporidia. Blackwell Scientific Publications, Oxford, 1,114 pp.

281. Perkins, F. O., Barta, J. R., Clopton, R. E., Pierce, M. A. & Upton, S. J. 2000. Phylum Apicomplexa Levine, 1970. pp. 190–369 in An Illustrated Guide to the Protozoa, 2nd ed., Vol. 1, Lee, J. J., Leedale, G. F. & Bradbury, P. (eds.). Society of Protozoologists, Lawrence, KS.

282. Telford, S. R. 1994. Plasmodia of reptiles. pp. 1–71 in Parasitic Protozoa, 2nd ed., Vol. 7, Kreier, J. P. (ed.). Academic Press, San Diego.

283. Poinar, Jr., G. O. 2005. *Plasmodium dominicana* n. sp. (Plasmodiidae: Haemospororida) from Tertiary Dominican amber. Systematic Parasitology 61: 47–52.

**284.** Baker, J. R. 1965. The evolution of parasitic protozoa. pp. 1–27 in Evolution of Parasites, Taylor, A. E. R. (ed.). Blackwell Scientific Publications, Oxford.

**285.** Hughes, A. L. & Piontkivska, H. 2003. Phylogeny of Trypanosomatidae and Bodonidae (Kinetoplastida) based on 18S rRNA: Evidence for paraphyly of *Trypanosoma* and six other genera. Molecular Biology and Evolution 20: 644–652.

**286.** Simpson, A. G. B., Stevens, J. R. & Lukes, J. 2006. The evolution and diversity of kinetoplastid flagellates. Trends in Parasitology 22: 168–174.

**287.** Schlein, Y. & Jacobson, R. L. 1998. Cellulase enzymes and the evolution of trypanosomatids. pp. 117–134 in Digging for Pathogens, Greenblatt, C. L. (ed.). Balaban Publications, Rehovot, Israel.

**288.** Lom, J. 1979. Biology of the trypanosomes and trypanoplasms of fish. pp. 269–337 in Biology of the Kinetoplastida, Vol. 2, Lumsden, W. H. R. & Evans, D. A. (eds.).Academic Press, London.

**289.** Lillegraven, J. A., Kraus, M. J. & Brown, T. M. 1979. Paleogeography of the world of the Mesozoic. pp. 277–308 in Mesozoic Mammals, Lillegraven, J. A., Kielan-Jaworowska, Z. & Clemens, W. A. (eds.). University of California Press, Berkeley.

**290.** Donnelly, T. W. 1992. Geological setting and tectonic history of Mesoamerica. pp.1–13 in Insects of Panama and Mesoamerica, Quintero, D. & Aiello, A. (eds.).Oxford University Press, New York.

**291.** Glick, P. A. 1939. The distribution of insects, spiders, and mites in the air. United States Department of Agriculture Technical Bulletin 673: 1–150.

**292.** Borror, D. J., Triplehorn, C. A. & Johnson, N. F. 1989. An Introduction to the Study of Insects, 6th ed. Saunders College Publishing, Philadelphia, 875 pp.

**293.** Swan, L. 1964. Beneficial Insects. Harper Row, New York, 429 pp.

**294.** Mikhailov, K. E. 1997. Eggs, eggshells, and nests. pp. 205–209 in Encyclopedia of Dinosaurs, Currie, P. T. & Padian, K. (eds.). Academic Press, San Diego.

**295.** Upchurch, P., Barrett, P. M. & Dodson, P. 2004. Sauropoda. pp. 259–322 in The Dinosauria, 2nd ed., Weishampel, D. B., Dodson, P. & Osmolska, H. (eds.). University of California Press, Berkeley.

**296.** Miller, D., Summers, J. & Sieber, S. 2004. Environmental versus genetic sex determination: A possible factor in dinosaur extinction? Fertility and Sterility 81: 954–964.

**297.** Price, P. W. 2002. Species interactions and the evolution of biodiversity. pp. 3–25 in Plant-Animal Interactions, Herrera, C. M. & Pellmyr, O. (eds.). Blackwell Scientific, Oxford.

**298.** Raup, D. M. 1991. Extinction. Bad Genes or Bad Luck? W. W. Norton and Co., New York, 210 pp.

**299.** Fastovsky, D. E. & Weishampel, D. B. 1996. The Evolution and Extinction of the Dinosaurs. Cambridge University Press, Cambridge, 460 pp.

**300.** Johnson, C. C., & Kauffman, E. G. 1996. Maastrichtian extinction patterns of Caribbean Province rudistids. pp. 231–273 in The Cretaceous-Tertiary Mass Extinctions: Biotic and Environmental Events, MacLeod, N. & Keller, G. (eds.). W.W. Norton & Co, New York.

**301.** Johnson, K. R. & Hickey, L. J. 1990. Megafloral changes across the Cretaceous/Tertiary boundary in the northern Great Plains and Rocky Mountains, USA. pp. 433–444 in Global Catastrophes in Earth History, Sharpton, V. L. & Ward, P. D. (eds.). Geological Society of America special paper 247, Boulder, CO.

**302.** Johnson, K. R. 1993. High latitude deciduous forests and the Cretaceous-Tertiary boundary in New Zealand. Abstracts, Geological Society of America annual meeting, Boston 25: A295.

**303.** Stromberg, C. A. E., Thompson, A., Arens, A. & Clemens, W. A. 1998. Precursors to the Cretaceous-Tertiary boundary event: Evidence for terrestrial environmental instability. University of California Museum of Paleontology 75/125 years anniversary symposium 28 Feb. 1998, http:// www.ucmp.berkeley.edu/museum/ 75th.

**304.** Fleming, R. F. 1985. Palynological observations of the Cretaceous/Tertiary boundary in the Raton Formation, New Mexico. Palynology 9: 242.

**305.** Nichols, D. J., Jarzen, D. M., Orth, C. S. & Oliver, P. O. 1986. Palynological and iridium anomalies at Cretaceous/Tertiary boundary, south-central Saskatchewan. Science 231: 714–717.

**306.** Worster, D. 1979. Dust Bowl: The Southern Plains in the 1930s. Oxford University Press, New York, 277 pp.

**307.** Schubert S. D., Suarez, M. J., Pegion, P. J., Koster, R. D. &

Bacmeister J. T. 2004. On the cause of the 1930s Dust Bowl. Science 303: 1855–1859.

308. Miller, Jr., G. T. 1988. Environmental Science, 2nd ed., Wadsworth Publishing, Belmont, CA, 407 pp.

309. Wilson, E. O. 1992. The Diversity of Life. The Belknap Press of Harvard University Press, Cambridge, 424 pp.

310. Whalley, P. 1988. Insect evolution during the extinction of the Dinosauria. Entomologica Generalis 13: 119–124.

311. Labandiera, C. C., Johnson, K. J. & Wilf, P. 2002. Impact of the terminal Cretaceous event on plant-insect associations. Proceedings of the National Academy of Sciences 99: 2061–2066.

312. Pike, E. M. 1994. Historical changes in insect community structure as indicated by Hexapods of Upper Cretaceous Alberta (Grassy Lake) amber. The Canadian Entomologist 126: 695–702.

313. Chitwood, B. G. & Chitwood, M. B. 1950. Introduction to Nematology. University Park Press, Baltimore, 334 pp.

314. Sohlenius, B. 1980. Abundance, biomass, and contribution to energy flow by soil nematodes in soil ecosystems. Oikos 34: 186–194.

315. Yuen, P. H. 1966. The nematode fauna of the regenerated woodland and grassland of broadwalk wilderness. Nematologica 12: 195–214.

316. Malakhov, V. V. 1994. Nematodes. Smithsonian Institution Press, Washington D.C., 286 pp.

317. Andrassy, I. 1983. A Taxonomic Review of the Suborder Rhabditina (Nematoda: Secernentia). Editions de l'office de la recherché scientific et technique outré-mer, Paris, 241 pp.

318. Poinar, Jr., G. O. 1990. Taxonomy and biology of Steinernematidae and Heterorhabditidae. pp. 23–61 in Entomopathogenic Nematodes in Biological Control, Gaugler, R. & Kaya, H. K. (eds.). CRC Press, Boca Raton, FL.

319. Jenkins, W. R. & Taylor, D. P. 1967. Plant Nematology. Reinhold Publishing, New York, 270 pp.

320. Jablonski, D. 1987. Mass extinctions: New answers, new questions. pp. 43–62 in The Last Extinction, Kaufman, L. & Mallory, K. (eds.). The MIT Press, Cambridge, MA.

321. Sullivan, R. M. 1998. The many myths of dinosaur extinction: Decoupling dinosaur extinction from the asteroid impact. pp.

58–59 in The Dinofest Symposium, Wolberg, D. I., Gittis, K. Carey, L. & Raynor, A. (eds.). Academy of Natural Sciences, Philadelphia.

**322.** Russell, D. A. 1984. Terminal Cretaceous extinctions of large reptiles. pp 383–384 in Catastrophes and Earth History, Berggren, W. A. & Van Couvering, J. A. (eds.). Princeton University Press, Princeton, NJ.

**323.** Russell, D. A. 1984. The gradual decline of dinosaurs—fact or fallacy? Nature 307: 360–361.

**324.** Dodson, P. 1990. Ceratopsidae. pp. 494–513 in The Dinosauria, Weishampel, D. B., Dodson, P. & Osmólska, H. (eds.). University of California Press, Berkeley.

**325.** Fassett, J. E., Zielinski, R. A. & Budahn, J. R. 2002. Dinosaurs that did not die: Evidence for Paleocene dinosaurs in the Ojo Alamo sandstone, San Juan Basin, New Mexico. pp. 307–336 in Catastrophic Events and Mass Extinctions: Impacts and Beyond, Koeberl, C. & MacLeod, K. G. (eds.). Geological Society of America Special Paper 356, Boulder, CO.

**326.** Benton, M. J. 1990. Scientific methodologies in collision. The history of the study of the extinction of the dinosaurs. Evolutionary Biology 24: 371–400.

**327.** Desowitz, R. S. 1991. The Malaria Capers. W. W. Norton & Co., New York, 288 pp.

**328.** Desjeux, P. & Alva, J. 2003. Leishmania/HIV co-infections: Epidemiology in Europe. Annals of Tropical Medicine and Parasitology 97 (Suppl. 1): 3–15.

**329.** Boots, M. & Sasaki, A. 2002. Parasite-driven extinction in spatially explicit host-parasite systems. The American Naturalist 159: 706–713.

**330.** Blaustein, A. R. & Dobson, A. 2006. A message from the frogs. Nature 439: 143–144.

**331.** Laurence, W. F., McDonald, K. R. & Speare, R. 1996. Epidemic disease and the catastrophic decline of Australian rain forest frogs. Conservation Biology 10: 406–413.

**332.** Van Riper, C., van Riper, S. G., Goff, M. L. & Laird, M. 1986. The epizootiology and ecological significance of malaria on the birds of Hawaii. Ecological Monographs 56: 327–344.

**333.** Cunningham, A. A. & Daszak, P. 1998. Extinction of a species of land snail due to infection with a microsporidian parasite. Conservation Biology 12: 1139–1141.

334. Cunningham, A. A. 2005. A walk on the wild side—emerging wildlife diseases. British Medical Journal 331: 1214–1215.

335. Perrins, C. 1979. Birds, Their Life, Their Ways, Their World. The Reader's Digest Association, New York, 416 pp.

336. Alvarez, L. W. 1980. Extraterrestrial cause for the Cretaceous-Tertiary extinction. Science 208: 1095–1108.

337. Poinar, Jr., G. O. & Buckley, R. 2006. Nematode (Nematoda: Mermithidae) and hairworm (Nematomorpha: Chordodidae) parasites in Early Cretaceous amber. Journal of Invertebrate Pathology 93: 36–41.

338. Poinar, Jr., G. O. & Szadziewski, R. 2006. *Corethrella andersoni* (Diptera: Corethrellidae), a new species from Lower Cretaceous Burmese amber. Proceedings of the Entomological Society of Washington 109: 155–159.

339. Courtillot, V. E. 1990. A volcanic eruption. Scientific American 256: 44–60.

340. Archibald, J. D. 2002. Dinosaur extinction: Changing views. pp. 99–106 in Dinosaurs: The Science Behind the Stories, Scotchmoor, J. G., Springer, D. A. & Breithaupt, B. H. (eds.). American Geological Institute, Alexandria, VA.

341. Sharpton, V. L. & Marin, L. E. 1997. The Cretaceous-Tertiary impact crater and the cosmic projectile that produced it. Annals of the New York Academy of Sciences 822: 353–380.

342. Fastovsky, D. E., Huang, Y., Hsu, J., Martin-McNaughton, J., Sheehan, P. M. & Weishampel, D. B. 2004. Shape of Mesozoic dinosaur richness. Geology 32: 877–880.

343. Archibald, J. D. 1996. Dinosaur Extinction and the End of an Era: What the Fossils Say. Columbia University Press, New York, 237 pp.

344. Poinar, Jr., G. O., Lambert, J. B. & Wu, Y. 2007. Araucarian source of fossiliferous Burmeses amber: Spectroscopic and anatomical evidence. Journal of the Botanical Research Institute of Texas 1: 449–455.

345. Poinar, Jr., G. O. & Buckley, R. 2007. Evidence of Mycoparasitism and Hypermycoparasitism in Early Cretaceous Amber. Mycological Research 111: 503–506.

346. Poinar, Jr., G. 2007. Early Cretaceous flagellates associated with fossil sand fly larvae in Burmese amber. Memórias do Instituto Oswaldo Cruz 102: 635–637.

**347.** Poinar, Jr., G. O., Kirejtshuk, A. G. & Buckley, R. 2008. *Pleuroceratos burmiticus* n. gen., n. sp. (Coleoptera: Silvanidae) from Early Cretaceous Burmese amber. Proceedings of the Entomological Society of Washington 110: (in press).

**348.** Poinar, Jr., G. O., Chambers, K. L. & Buckley, R. 2007. *Eoëpigynia burmensis* gen. and sp. nov., an Early Cretaceous eudicot flower (Angiospermae) in Burmese amber. Journal of the Botanical Research Institute of Texas 1: 91–96.

**349.** Poinar, Jr., G. O., Marshall, C. & Buckley, R. 2007. One hundred million years of chemical warfare by insects. Journal of Chemical Ecology 33: 1663–1665.

# Index

Note: Figures are in italics; color plates are in bold